WHY IS IT?

QUESTION & ANSWER ENCYCLOPEDIA

WHY IS IT?

General Editor   Lesley Firth

KINGFISHER BOOKS

First published in 1983 by Kingfisher Books Limited
Elsley Court, 20 – 22 Great Titchfield Street
London W1P 7AD

BRITISH LIBRARY CATALOGUING IN PUBLICATION DATA

Why is it?—(Question and answer encyclopedia)
   1, Science—Dictionaries—Juvenile literature
   I. Firth, Lesley   II. Series
   503'.21   Q123
   ISBN 0-86272-071-0
   D. L. TO: 793-1983
Phototypeset by Southern Positives and Negatives
(SPAN) Lingfield, Surrey

Colour separations by Newsele Litho Ltd,
Milan, Italy

Printed in Spain by Artes Gráficas Toledo, S.A.

Authors

Neil Ardley
Mark Lambert
Christopher Maynard
James Muirden
Christopher Pick
Brian Williams
Jill Wright

Artists

Bob Bampton/The Garden Studio
Norma Burgin/John Martin & Artists
Paul Crompton/John Martin & Artists
Dave Etchell & John Ridyard
Oliver Frey/Temple Art Agency
Jerry Hoare/John Martin & Artists
Ron Jobson/Tudor Art Agency
Tony Payne
Mike Roffe
Mike Saunders/Jillian Burgess
Charlotte Styles
Tammy Wong/John Martin & Artists

CONTENTS

PLANTS AND ANIMALS

What makes living things special?	10
Why are plants green?	10
Why do plants need water?	10
Who feeds on what?	11
Why do some plants lose their leaves in autumn?	11
What are flowers for?	12
Why are insects necessary to plants?	12
Why do flowers have bright colours?	12
How much pollen?	13
Which flowers are not brightly coloured?	13
Why do some plants have no flowers?	13
Which fruits explode?	14
Which fruits have wings and parachutes?	14
Which fruits are carried by animals?	14
Why does the lords and ladies plant trap insects?	15
Which plants stink?	15
What are bulbs, corms and rhizomes?	15
Why do stems grow up and roots grow down?	16
Why do plants bend towards the light?	16
Do plants feel?	16
Why do some plants eat animals?	17
Which plant sets a vicious trap to catch animals?	17
Which plants catch animals in underwater bladders?	17
Why do jellyfish sting?	18
What do limpets eat?	18
Why does a spider build a web?	18
Why do wasps sting?	19
Why do leafcutter ants cut leaves?	19
Which insects drink through straws?	19
Which fishes kill with electricity?	20
How do birds feed?	20
Do vampires suck blood?	20
Do bears eat honey?	21
Why does the elephant have a trunk?	21
Why does a giraffe have a long neck?	21
Which fishes 'see' with electricity?	22
Why do some animals have no eyes?	22
How does a worm hold on?	22
Why does a snake's tongue flicker in and out?	23
How does a fly walk upside down?	23
How high can a flea jump?	23

Why do some fish clean others?	24
Why do animals hibernate?	24
Why do a fish and a shrimp share a home?	24
Where does a lungfish go when its stream dries up?	25
Who feeds a cuckoo?	25
Why do some ants live in thorn bushes?	25
How well can a moth smell?	26
Which insect lays 1000 eggs a day?	26
Why do bees swarm?	26
Which males must beware?	27
Why do sticklebacks change colour?	27
Why do birds sing?	27
Why do grebes dance?	28
Which bird is the greatest wooer?	28
Which birds build apartment blocks?	28
Why do marsupials have pouches?	29
Why do beavers build dams?	29
Are all spiders poisonous?	30
Why does an antelope leap in the air?	30
Why do bees sting?	30
Do flying fishes fly?	31
Why do hedgehogs have spines?	31
Why are some snakes dangerous?	31
How does an animal become invisible?	32
When is a twig not a twig?	32
Why are some animals brightly coloured?	32
Why does a chameleon change colour?	33
Why do some animals have false eyes?	33
Why does a squid squirt ink?	33

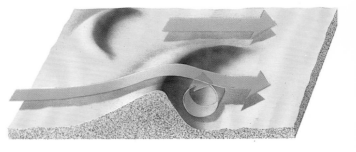

OURSELVES

Why do the police use fingerprints? 34
What do a black skin, a suntan and freckles
   have in common? 34
Why do you shiver when you are cold? 35
What are goose pimples? 35
Why do you sweat? 35

Why do some people have blue eyes and
   others brown eyes? 36
Why do bruises go black and blue? 36
Why does blood clot? 36
Why does hair go grey? 37
What makes us cry? 37
What protects your eyes? 37

Why do you need vitamins and minerals? 38
Does it matter what you eat? 38
Does it matter how much you eat? 38
Why do we need exercise? 39
What is your body made of? 39
Why do you need rest? 39

Why do you dream? 40
Why is your temperature taken when you are ill? 40
Why should you brush your teeth? 40
Why is smoking bad for you? 41
Is alcohol good or bad? 41
Why are vaccinations necessary? 41

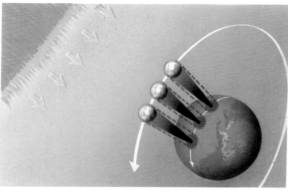

SCIENCE

Why can you see yourself in a mirror? 42
Why do things fall to the ground? 42
How does glue stick things together? 43
Why does soap make bubbles? 43
Why does paper tear easily? 43

Why do rubber balls bounce? 44
Why do things go rusty? 44
Why does a balloon float in the air? 44
Why does a spinning top not fall over? 45
Why do wet things shine? 45
Why are there only ten numbers with single
   figures? 45

Why does sugar dissolve in water? 46
Why does a boat float on water? 46
Why does water flow more easily than syrup? 46
Why does water freeze in winter? 47
Why does a puddle dry up in sunshine? 47
Why is ice slippery? 47

Why does metal feel cold? 48
Why does wood burn but not iron? 48
Why do clothes keep you warm? 48
Why is a flame hot? 49
Why do matches light when you strike them? 49
Why does an electric fire glow red? 49

TRANSPORT

Why are there different types of tyre? 50
Why do car designs differ? 50
Why do we wear seat belts? 51
Why are traffic signals necessary? 51
Why do bicycle wheels have spokes? 51

Why do city railways run underground? 52
Why do railways use bridges and tunnels? 52
Why do railway tracks need regular
   maintenance? 52
Why are railway signals so important? 53
Why do trains run on rails? 53
Why are some railways electrified? 53

Why were most early aircraft biplanes? 54
Why do some aircraft have propellers? 54
Why are passenger airships no longer used? 55
Why do hot-air balloons fly? 55
Why do some aircraft have jet engines? 55

Why do yachts have different types of sail? 56
Why do boats have keels? 56
Why are there different types of merchant
   ship? 56
Why do ships have load-lines? 57
Why were submarines invented? 57
Why do canals have locks? 57

OUT IN SPACE

Why do the planets revolve round the Sun?	58
Why is there life on the Earth?	58
Why can't we see other solar systems?	58
Why is there no weight in space?	59
What is the difference between a planet and a moon?	59
Why does the Moon keep the same face towards the Earth?	60
Why does the Moon wax and wane?	60
Why do eclipses of the Moon occur?	60
Why is the Moon covered with craters?	61
Why is the New Moon invisible?	61
Why is there no life on the Moon?	61
Why is Mars a dead world?	62
Why is Mars red?	62
Which planet spins the fastest?	62
Why is Venus so hot?	63
Why is it difficult to see Mercury?	63
Why does Saturn have rings?	64
Why are the giant planets gaseous?	64
Why were the asteroids formed?	64
Why do many comets appear unexpectedly?	65
Why does a comet have a tail?	65
Why do meteors occur?	65
Why must you never look at the Sun?	66
Why do eclipses of the Sun occur?	66
Why is the setting Sun red?	66
Why does the Sun keep shining?	67
Why are there spots on the Sun?	67
How big is the Sun?	67
Why are stars invisible in the daytime?	68
Why do stars twinkle?	68
Why are some stars brighter than others?	68
How are stars formed?	69
How do stars die?	69
Why does the Milky Way look patchy?	69

PLANET EARTH

Why does the Earth look so blue from space?	70
Why does a compass needle point north?	70
Why does the Sun rise in the east?	71
Why do we have day and night?	71
Why do we have seasons?	71
Why did the dinosaurs die out so suddenly?	72
Why are fossils important to geologists?	72
Why do some rocks contain fossils and not others?	72
Why are there coal seams in the Antarctic?	73
Why do the continents move about?	73
Why have some seas disappeared?	73
Why do volcanoes erupt?	74
Why do earthquakes and volcanoes occur only in some parts of the world?	74
Why do earthquakes occur?	74
Why does a geyser spout hot water and steam?	75
Why are there great mountain ranges?	75
Why does it get colder as you climb a mountain?	75

Why is sea water salty?	76
Why do ocean currents occur?	76
Why do the tides rise and fall?	76
Why do waves break?	77
Why do ripple marks appear on sandy beaches?	77
Why do seas sometimes erode the land?	
Why are cloudy nights warmer than clear star-lit nights?	78
Why do winds blow?	78
Why do weather systems have a circular pattern of winds?	78
Why do clouds form?	79
What makes it rain?	79
What causes fog?	79
Why does thunder always follow lightning?	80
Why does lightning often strike trees?	80
Why do we sometimes get hail?	80
Why is a snowflake made up of crystals?	81
Why do hurricanes cause so much damage?	81
Why do we see rainbows?	81

Why do river meanders change their shape? 82
Why do some rivers have deltas? 82
Why do lakes sometimes disappear? 82
Why do glaciers appear? 83
How do glaciers move? 83
Why have ice-sheets altered the landscape? 83

Why are some soils fertile? 84
Why is limestone different from other rocks? 84
Why do underground caves occur? 84
Why does soil erosion occur? 85
Why do sand dunes move across deserts? 85
Why are some deserts getting bigger? 85

THE PAST

Why do human beings walk upright? 86
Why did people start to live in towns? 86
Why were the pyramids built? 87
Why did the Greeks build a wooden horse? 87
Why did Hannibal fight a war with the Romans? 87

Why were the early Christians persecuted? 88
Why did the Chinese build the Great Wall? 88
Why were the Vikings bold sea voyagers? 88
Why were medieval castles built? 89
Why did the Crusaders fight the Saracens? 89
Why did Luther disagree with the Church? 89

Why were the first Americans called Indians? 90
Why did Charles I of Spain boast: 'In my empire the sun never sets'? 90
Why did people begin to work in factories? 90
What caused the French Revolution? 91
Why was Italy unified? 91
Why did European nations colonize Africa? 91

Why did Russia have a revolution in 1917? 92
Why was World War I so costly? 92
Why did the Chinese make the Long March? 92
Why was the United Nations set up? 93
Why was the state of Israel founded? 93
Why was the Berlin Wall built? 93

HOW PEOPLE LIVE

Why are some babies baptized? 94
Why do people get married? 94
Why do we have funerals? 95
Why do Christians take Communion? 95
Why do Christians go to church on Sundays? 95

Why do people celebrate the New Year? 96
Why do Christians observe Lent? 96
Why is Christmas celebrated? 97
Why do we have Christmas trees? 97
Why do Christians celebrate Easter? 97

Why do Muslims face Mecca to pray? 98
Why do people make religious pilgrimages? 98
Why do the Jewish people celebrate Passover? 98
Why do Hindus worship many gods and goddesses? 99
Why do Hindus and Buddhists meditate? 99
Why do mosques have minarets? 99

Why do some Muslim women wear veils? 100
Why do the Scots wear tartans? 100
Why do Sikhs wear turbans? 100
Why do people use cosmetics? 101
Why do people wear uniforms? 101
Why do people like jewellery? 101

Why do we use soap? 102
Why do we use perfumes? 102
Why do we wear hats? 102
Why do we wear shoes? 103
Why are some people tattooed? 103
Why do hairstyles vary so much? 103

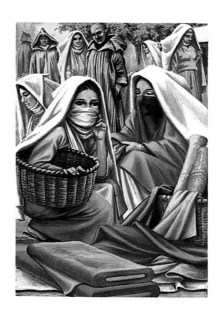

ARTS, SPORT AND ENTERTAINMENT

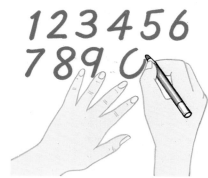

Why do human beings make music? 104
Why did Haydn write the 'Farewell'
  Symphony? 104
Why do ballerinas wear blocked shoes? 105
Why does an orchestra need a conductor? 105
Why does an oboe play before the rest of an
  orchestra? 105

Why do actors wear make-up? 106
Why do plays and films have directors? 106
Why was Odysseus tied to a mast? 106
Why did Michelangelo take years to do one
  painting? 107
Why is the Mona Lisa smiling? 107
Why are some paintings called abstract? 107

Why are there rings in the circus? 108
Why do circuses have clowns? 108
Why do tightrope walkers sometimes carry
  long poles? 108
Why do magicians use props? 109
Why do people believe in magic? 109
Why are there kings and queens in chess? 109

Why do snakes 'dance' for snake charmers? 110
Where did puppets come from? 110
Why is bullfighting found mainly in Spain? 110
Why do people read horoscopes? 111
Why are there 52 cards in a pack? 111
How do ventriloquists 'throw' their voices? 111

Why do runners start from different positions
  in some track races? 113
Why can some races be won by walking? 113

Why is discus-throwing still a sport? 114
Why are pole-vaulters not hurt when they
  land? 114
Why are there barriers in some races? 114
Why is gymnastics an Olympic sport? 115
Why is real tennis different from tennis? 115
Why do most golfers have handicaps? 115

Why is a yellow jersey worn during the *Tour
  de France*? 116
Why do racing drivers wear flame-proof
  clothes? 116
Why are coloured belts worn in karate? 116
Why are Sumo wrestlers giants? 117
Why do ice-hockey players wear masks and pads? 117
Why do boxers wear gloves? 117

Why are some styles of swimming faster than
  others? 118

Why are the Olympic Games held every four
  years? 112
Why are professional sportsmen different
  from amateurs? 112
Why are very long races known as marathon
  events? 112
Why do runners use starting blocks? 113

Why do surfboards travel so swiftly? 118
Why is soccer so widely played? 118
Why are cameras so important in track races? 119
Why are skis so long? 119
Why do rowing teams often carry passengers? 119

INDEX 120

PLANTS AND ANIMALS

▶ WHAT MAKES LIVING THINGS SPECIAL?

Living things range from tiny, single-celled organisms to huge trees and mammals. But they all have seven special features in common.

All living things need food. They use some of it for growing and some is used to create energy. This is done by breaking it down in oxygen taken from the air or water in which they live. This process is called respiration. Waste food is got rid of, or excreted. So four features of living things are feeding, growth, respiration and excretion.

Two more features are movement and sensitivity. These are most obvious in animals, but are also shown by plants. The last feature is that all living things reproduce themselves.

▼ WHY ARE PLANTS GREEN?

Plants make their own food. To help them do this they have large amounts of a green pigment called chlorophyll in their cells. This pigment gives plants their green appearance.

Plants make food by a process known as photosynthesis. This word is made up of two parts: 'photo', meaning 'light', and 'synthesis', meaning 'building up'. Using water and carbon dioxide as raw materials, they build up food sugar in their cells. They get the energy for this from the Sun in the form of light. And the green pigment chlorophyll is essential in photosynthesis. Without it plants could not harness the Sun's light energy. So the food-producing parts of plants contain chlorophyll.

▶ WHY DO PLANTS NEED WATER?

Without water plants wilt and die. They lose water through their leaves. So to replace this water, they need a constant supply from the roots.

Over 90 per cent of a plant is water. This may seem like a lot, but it is vital to the plant for several reasons. First, all the chemical reactions that take place in a cell can occur only in water. And the process of photosynthesis actually

uses up water. Second, it helps keep each cell rigid. If the cells do not have enough water, they become limp and the plant wilts.

The plant loses water through tiny pores called stomata on the undersides of its leaves. This process is known as transpiration. At the same time water is drawn up into the roots from the soil. Transpiration has two uses. Water evaporating from the leaves helps to keep the plant cool. The flow of water up the stem also brings with it vital minerals from the soil.

Healthy plant

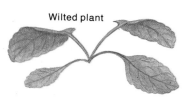
Wilted plant

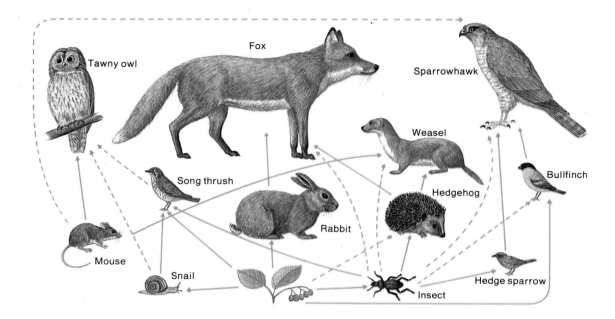

▲ WHO FEEDS ON WHAT?

Every animal and plant eats, or is eaten by, other living things. Together, a small group of living things form a food chain. Plants are eaten by plant-eating animals, which in turn are eaten by meat-eating animals. These meat-eaters may also be eaten.

A single food chain consists of a plant, a herbivore (plant-eating animal) and one or more carnivores (meat-eaters). However, a food chain rarely exists by itself. Most animals eat a variety of foods and may themselves be eaten by several predators. As a result they form part of a more complicated food web.

The illustration shows some of the animals that may be involved in a hedgerow community. The usual food eaten is shown by a solid line. Occasional food is shown by a dotted line. Plant-eaters, such as rabbits, voles, snails and insects feed on the leaves of plants. Some small mammals, birds and insects feed on the fruits and seeds. In turn, all these animals are preyed on by larger animals. At the top of the food web are the large predators, such as foxes, hawks and owls.

The chain does not really end there. Dead animals and plants are cleaned up by soil-dwellers such as earthworms, ground beetles and bacteria.

▶ WHY DO SOME PLANTS LOSE THEIR LEAVES IN AUTUMN?

A large number of trees shed their leaves in order to survive the cold during winter. They lie dormant for several months, avoiding water loss and frost damage.

In temperate regions of the world the soil is too cold during the winter months for most broadleaved trees to take up enough water. So to avoid losing water during the

Beech leaf (broadleaved)

Beech leaf in autumn

Pine needles (evergreen)

winter, these trees (known as deciduous trees) shed their leaves in autumn. Before they do so, however, the precious green chlorophyll is withdrawn. This leaves other pigments, such as yellows, oranges and reds, which give autumn leaves their familiar brilliant colours.

The evergreen trees of temperate regions are mostly conifers. They live in areas where water is scarce. They have thin, needle-like leaves that keep water loss to a minimum. So they can keep their leaves during the winter.

▶ WHAT ARE FLOWERS FOR?

Flowers produce seeds that can grow into new plants. They often have brightly coloured petals, which surround the male and female organs.

The parts of a flower all have important functions. They are held on a swollen base called the receptacle. The outer parts of the flower are green, leaf-like structures called sepals. They cover and protect the bud before the flower opens out. Inside the sepals are the petals. Brightly coloured petals attract insects or other animals to the flower.

The male organs are called stamens. Each one consists of an anther (four pollen-containing sacs) on the end of a long stalk or filament. The female organs are called carpels. Each one consists of an ovary and a pollen-receiving surface, or stigma, which is often on the end of a stalk, or style. The ovary contains one or more ovules, each of which has a female egg cell, or ovum, inside.

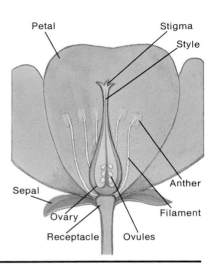

▲ WHY DO FLOWERS HAVE BRIGHT COLOURS?

Transferring pollen from an anther to a stigma is called pollination. In many cases this is done by animals. To attract animals, flowers have bright colours and often produce nectar.

Several kinds of animal pollinate flowers, but the chief pollinators are insects which are attracted to the flowers by their colours and scents. Often there are lines and markings on the petals to

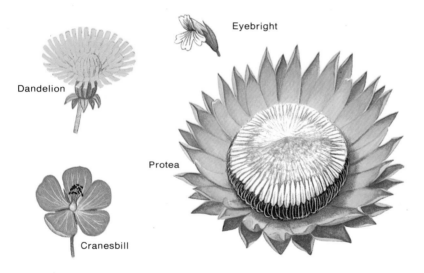

Eyebright

Dandelion

Protea

Cranesbill

guide insects towards the nectar (for example eyebright and cranesbill). Insect-pollinated flowers are usually blue, yellow, pink or white. However, the colours we see are not the same that insects see. To a bee, for example, red appears grey. But a bee can detect ultraviolet light. So a white or yellow flower probably looks blue to a bee.

In tropical and subtropical regions birds, bats, rodents and marsupials pollinate flowers. These flowers are often very brightly coloured. Protea, for example, is bird-pollinated.

◀ WHY ARE INSECTS NECESSARY TO PLANTS?

Insects are useful to plants because they carry pollen from one flower to another. This kind of pollination produces the strongest plants.

Transferring pollen from an anther to a stigma in the same flower is called self-pollination. Transferring pollen from one flower to another is called cross-pollination. It is better for plants to be cross-pollinated, because this results in healthier offspring. Self-pollination does happen, but usually only if cross-pollination has failed.

Many insect-pollinated flowers have ingenious ways of preventing self-pollination. Sometimes the stamens and carpels ripen at different times. Some plants, such as the primrose, have more than one kind of flower in which the stamens and stigmas are in different places. Pea flowers have flaps that prevent insects from brushing pollen collected from a flower onto the stigma of the same flower.

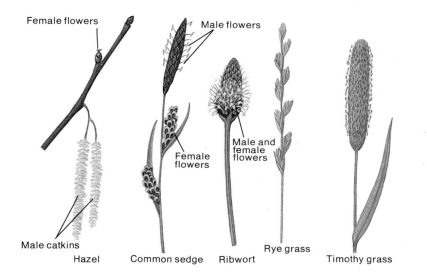

Female flowers

Male flowers

Female flowers

Male and female flowers

Rye grass

Male catkins

Hazel   Common sedge   Ribwort   Timothy grass

Wind pollination leaves a lot to chance. To make sure some pollen lands on the right stigmas, the anthers have to produce huge amounts.

A single rye flower produces about 50,000 grains of pollen and one silver birch catkin sheds over five million grains. Wind-pollinated plants produce their pollen in large stamens that hang out of the flowers. When they are ripe the anthers split open, shedding clouds of pollen into the air. Most of it is wasted, however. Only a small amount lands on the stigmas of female flowers. To increase their chances of catching pollen, female flowers often have large, feathery stigmas that act like nets.

In spite of the waste, wind pollination is very effective. Pollen is easily spread among the tall stems of grass plants that cover the ground. Wind-pollinated trees shed their pollen early in the year. At this time there are not many leaves to obstruct the pollen grains.

▲ WHICH FLOWERS ARE NOT BRIGHTLY COLOURED?

A number of plants are pollinated by wind. Such plants do not need large showy flowers to attract insects. Instead they have small flowers with tiny petals or none at all.

The largest group of wind-pollinated plants is the grass family. There are about 10,000 species and these include the cereals we grow for food, such as wheat, rice and maize. Grass flowers have no petals, but are protected by tiny leaf-like structures.

Other wind-pollinated plants include sedges, rushes and plantains. Some trees and shrubs are also pollinated by wind. These include hazel, birches, alders and willows. To ensure cross-pollination, these trees produce separate male and female flowers. Willows even have separate male and female trees. The male flowers are produced in bunches, known as catkins. The female flowers are produced in smaller catkins or cone-like structures.

► WHY DO SOME PLANTS HAVE NO FLOWERS?

Mosses and ferns do not produce seeds. Instead they reproduce by means of spores.

A moss plant develops sex cells amongst the leaves of its small shoots. The male cells reach female cells by swimming through the thin film of water on the leaves.

However, the fertilized female cell does not develop either into a seed or directly into a new moss plant. Instead it develops into a stalked capsule, which contains thousands of tiny spores. When the capsule is ripe, the spores are released. Each one is capable of producing at least one new moss plant. The spores of ferns develop on the leaves of the adult plant. When it lands on the ground, a spore develops into a tiny heart-shaped structure called a prothallus. This produces male and female cells. Only the tiny prothallus has to be wet. The adult fern can live in much drier conditions.

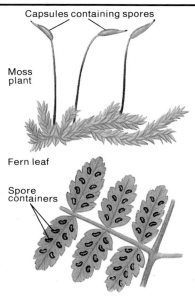

Capsules containing spores

Moss plant

Fern leaf

Spore containers

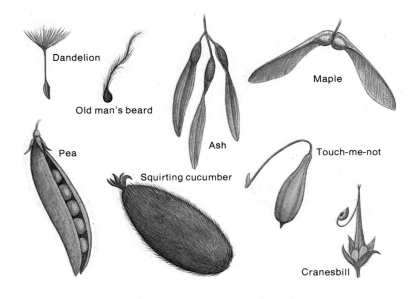

Dandelion

Old man's beard

Maple

Ash

Pea

Touch-me-not

Squirting cucumber

Cranesbill

Using the wind is a useful way of spreading seeds. Some seeds are light enough by themselves. Others are helped by wings or parachute-like tufts of hair.

Many seeds are dispersed by wind, and a number of them have specially designed fruits. The fruits of dandelions, thistles and many other members of the daisy family are crowned with hairy plumes that act as parachutes. The seeds of willowherbs and willows also have plumes like this. The fruits of old man's beard have hairy 'tails' that help the wind to carry them through the air.

The most spectacular winged fruits are found on trees. A birch fruit has a pair of small wings with the seed in between. An ash fruit forms a long, aerodynamic wing with the seed at one end. Maple trees have two-seeded, double-winged fruits that spin like helicopter blades as they fall to the ground.

▲ WHICH FRUITS EXPLODE?

One way of dispersing seeds is to hurl them away from the parent plant. Several plants have fruits that burst violently.

Most exploding fruits are capsules, pods or similar structures. They gradually lose water and, when they are dry, they split apart, throwing out the seeds with some force. For example, when the fruits of cranesbills and other geraniums break open, the seeds are hurled out on the ends of spring-like fibres.

Laburnum and many other members of the pea family produce their seeds in long pods. When a laburnum pod dries out it breaks open with a loud pop. Both halves twist suddenly, flinging the seeds out.

Some exploding fruits swell up with water. Touch-me-not balsam fruits become so full that they explode at the slightest touch. The squirting cucumber shoots out its seeds in a jet of water.

▶ WHICH FRUITS ARE CARRIED BY ANIMALS?

Many plants use animals as seed-carriers. Animal-dispersed seeds include those found in juicy fruits and hooked fruits.

Animals may carry fruits a long way from the parent plant. Juicy fruits, such as those of blackberry and hawthorn, provide birds and small rodents with tempting meals. The tough seeds are eaten with the fruit, but they pass through unharmed.

Some plants produce fruits

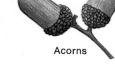

Burdock

Acorns

Goosegrass

Blackberries

with tiny hooks on them. Goosegrass and burdock are common examples.

Some animals disperse fruits by forgetting about them! Squirrels collect and store away oak acorns and hazel nuts. During the winter they often forget about some of their food stores.

Sometimes even insects disperse seeds. Snowdrop and sweet violet seeds are carried away by ants, who are rewarded by juice from an oily knob on each seed.

▼ WHY DOES THE LORDS AND LADIES PLANT TRAP INSECTS?

The lords and ladies plant makes sure that it is cross-pollinated. Flies are attracted by the rotten smell and fall inside the sheath. They pollinate the female flowers. Then they are showered with pollen and released.

Flies investigating the foul-smelling spike, or spadix, of the lords and ladies plant land on the inside of the green

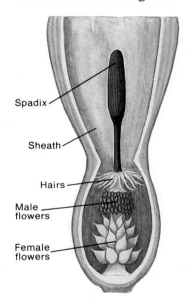

Spadix

Sheath

Hairs

Male flowers

Female flowers

sheath. This has a slippery surface and the flies slide down past a fringe of stiff, downward-pointing hairs.

The flies remain at the bottom of the trap all night. First they crawl over the female flowers, pollinating them with pollen brought from other plants. The stigmas then wither and the stamens of the male flowers ripen and burst, scattering pollen over the flies. Next morning, the slippery surface inside the sheath falls away, the downward-pointing hairs wither, and the flies escape.

▼ WHAT ARE BULBS, CORMS AND RHIZOMES?

Some plants can produce new plants without using flowers. They grow the new plants from their leaves or stems. Bulbs, corms and rhizomes are all kinds of stem used to produce new plants.

A bulb is a very short stem with closely-packed, fleshy food-storing leaves. New bulbs are formed in between the base of the leaves.

A corm is a swollen food-

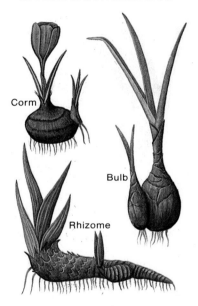

Corm

Bulb

Rhizome

storing stem. A new corm develops at the base of the flowering shoot, above the old corm, which dies.

Rhizomes are stems that grow horizontally under-ground, putting up shoots at intervals. Some plants, such as couch grass, have long, thin rhizomes. Others, such as irises, have thick rhizomes that act as food stores. Some plants, like the potato, have long thin rhizomes that swell in certain places into large tubers. These act as food stores. New shoots develop from buds on the tubers.

▼ WHICH PLANTS STINK?

Some plants attract insect pollinators in an unusual way. Instead of smelling sweetly of scent, they stink like rotting meat. This attracts pollinating flies.

One of the best known stinking flowers is that of the Malayan plant *Rafflesia*. It is also the world's largest flower and flies are attracted by its smell and its dark red-brown petals.

The African stapeliads also produce stinking flowers,

Rafflesia

known as carrion flowers. Blow-flies are sometimes so convinced that a flower really is rotting meat that they lay their eggs on it. But when the maggots hatch out they die, because there is actually nothing for them to feed on.

The lords and ladies plant also attracts flies by the smell given out by the spike that sticks out of the top. One of the lords and ladies' tropical relatives stands two metres high and gives out a smell like rotten fish.

15

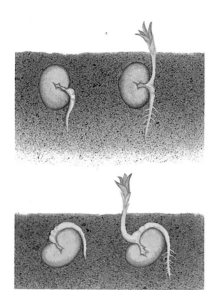

◀ WHY DO STEMS GROW UP AND ROOTS GROW DOWN?

Plants respond to the Earth's gravity. They can 'sense' which way is up and which way is down, so that their roots grow down into the soil and their stems grow up towards the light.

When a seed begins to germinate, it does not matter which way up it is lying. The root will always grow downwards and the stem will grow upwards. In the same way, if you turn a seedling upside down, the root will soon curl downwards again and the stem will turn upwards.

This is important to the plant. The root has to grow down into the soil to obtain water. The leaves must be in sunlight so that photosynthesis can take place.

The response of a plant to gravity is called geotropism. It is controlled by certain plant hormones. When a root or stem tip is growing in the wrong direction, the hormones cause it to change direction by slowing down the rate of growth on one side.

▶ WHY DO PLANTS BEND TOWARDS THE LIGHT?

Plants need light for photosynthesis. They can 'sense' where light is coming from and will always grow towards it.

You can test the effect of light on plants yourself. Take a tray of young plants that have been grown in the open. Place them indoors, near a window. All the plants will soon bend towards the light.

The ability to detect light is important for a plant, as light

is essential for photosynthesis. In some natural surroundings, such as the base of a hedge, plants need to be able to grow outwards, away from the shady area, to get enough light.

The response of plants to light is called phototropism. It is controlled by hormones. If a plant is illuminated from one side only, it grows faster on the shaded side and therefore bends towards the light.

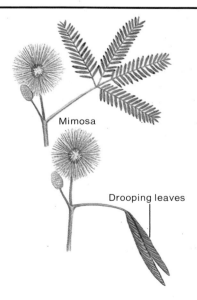

Mimosa

Drooping leaves

◀ DO PLANTS FEEL?

Plants react to light and gravity. They also react to touch and sound. However, they do not feel pain or emotion as people do.

Plants are sensitive to certain stimuli. They respond to light and gravity and many plants react to touch. The leaves of the plant *Mimosa pudica* droop suddenly if disturbed.

It is possible to cause different amounts of reaction in *Mimosa*. A gentle touch may cause only one leaf to droop. But a large disturbance can affect the whole plant. This shows that a message is being transmitted through the plant.

Plants have also been shown to grow faster when there is sound, such as music, around them. Some people even claim that plants can communicate by telepathy and that they can feel pain and emotion. Although there is no doubt that plants can feel to a limited extent, they certainly do not feel in the same way that animals do.

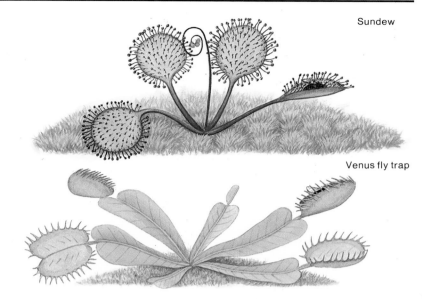

Sundew

Venus fly trap

▶ WHY DO SOME PLANTS EAT ANIMALS?

Plants that live in poor soil conditions may have difficulty in getting all the minerals they need. Several kinds of plant have overcome this problem by devising ways of killing and eating animals. Sundews are insect-eating plants.

In swamps and bogs, plants may not be able to take up enough minerals, such as nitrogen. Sundews are peat-bog plants that solve this problem by trapping insects.

The leaves of a sundew are covered in tentacles, each of which has a drop of sticky fluid on the end. An insect landing on a leaf becomes glued to it. As it struggles, it sets off the other tentacles into action. They bend over towards the insect, which is soon trapped. The tentacles then produce digestive enzymes, which break down the soft parts of the insect's body. When the insect is digested, the tentacles open up and the remains are blown away.

▲ WHICH PLANT SETS A VICIOUS TRAP TO CATCH ANIMALS?

The Venus fly trap has fearsome-looking traps on the ends of its leaves. When an unwary insect lands inside a trap, it swiftly closes and the insect is digested.

Venus fly traps are found in the bogs of California and Florida. A trap consists of two lobes, each fringed with spikes. Inside the spikes are scent glands that attract

insects. On each lobe there are three trigger hairs.

When an animal (usually an insect) lands on a lobe, it may touch a trigger hair. If it then touches another trigger hair (or the same hair again) within a fairly short time, the lobes close within a quarter of a second. The interlocking spikes prevent the insect from escaping.

The lobes then push inwards, crushing the insect between them. The insect's body is broken down by acid and absorbed by the plant.

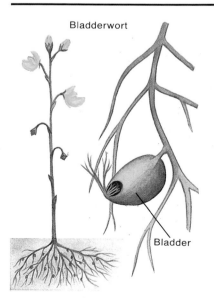

Bladderwort

Bladder

◀ WHICH PLANTS CATCH ANIMALS IN UNDERWATER BLADDERS?

Bladderworts catch their prey by using suction traps. Their leaves bear a number of small bladders. A tiny animal that comes too close to a bladder is sucked in and dies.

Most bladderworts are water plants that drift near the surface of ponds and lakes. Only the flower spikes are above the surface. Each bladder is flask-shaped, with a

trap-door surrounded by bristles at one end. It looks rather like a small crustacean. This seems to attract tiny crustaceans, such as the water flea *Daphnia*, which may be in search of protection.

The trap is ready to work when water is pumped out, creating a partial vacuum inside. When an animal touches a trigger hair, the door flies open and water rushes in, carrying the animal with it. The animal dies and decays. Its remains are absorbed by star-shaped glands inside the bladder.

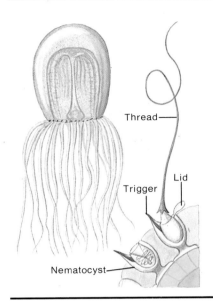

◀ WHY DO JELLYFISH STING?

All animals have their own ways of feeding. Jellyfish and their relatives, such as sea anemones, catch their prey with their tentacles. These are often armed with poisonous stings.

The tentacles of a jellyfish are covered in tiny cells called cnidoblasts. Each one produces a structure called a nematocyst. This consists of a hollow thread, a lid and a trigger hair. When a prey animal comes close, it stimulates the trigger hair. The lid flies open and the thread shoots out.

There are several different kinds of nematocyst. Some have long, barbed threads that become entangled with the prey. Others have sticky threads. And some have shorter, tougher threads that can pierce the victim. Poison is then injected through the hollow thread and the prey is paralyzed.

Once the prey is caught, the jellyfish pushes it into its mouth with its tentacles.

▶ WHAT DO LIMPETS EAT?

Limpets and many other shellfish feed on tiny seaweeds, or algae. They wander over the rocks, using a rasp-like 'tongue' to scrape off the algae as they go.

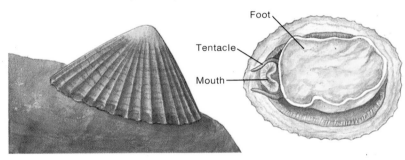

At low tide limpets can be found clinging tightly to the rocks. To make a perfect airtight fit, the limpet grinds a shallow cavity into the rock. The shell fits exactly into this cavity, and the limpet returns to this spot each time the tide goes out.

At high tide, when the limpet is underwater, it leaves its base and wanders over the rock looking for food. It moves by a rippling action of its large foot. Inside its mouth is a small sac that contains a kind of rasping 'tongue' called a radula. It is a long membrane attached to the base of the sac and it has many rows of curved teeth. When the limpet feeds, it pushes the radula out of its mouth and scrapes off the tiny algae that live on the rock.

◀ WHY DOES A SPIDER BUILD A WEB?

Many spiders catch their prey in silken webs. An insect that blunders into one of these webs finds it difficult to escape.

There are two main kinds of web-building spider. Some build shapeless sheetwebs, like the familiar cobweb of the house spider. Others, such as the garden spider, build beautiful, cartwheel-like orbwebs.

A sheetweb spider builds 'trip-wires' into its web and hides, often in a silk tunnel, near the edge. An insect that lands on the web stumbles and loses its balance. Before it can fly away again, the spider rushes out and seizes it.

The webs of orbweb spiders are sticky. A small insect that flies into one of these cannot escape. Its struggles alert the spider.

Most spiders catch insects in their webs. But some large tropical spiders build webs that can trap small birds.

◄ WHY DO WASPS STING?

Wasps use their stings to kill other insects, which they feed to their grubs. They only sting people when they are annoyed or become trapped.

Adult wasps feed on nectar, fruit and tree sap. But their grubs, or larvae, feed on the bodies of other insects.

In a colony of social wasps the adult workers go out hunting. When an insect is caught it is killed by the wasp's sting. Then it is taken back to the nest and chewed up. Its juices are fed to the larvae.

Solitary wasps, such as the sand wasp shown here, also hunt and sting live food for their larvae. The prey is paralyzed and dragged back to the wasp's burrow. Then the wasp lays an egg. When the larva hatches out, it feeds on the prey's body. Most solitary wasps hunt other insects, such as caterpillars. But some tropical wasps prey on large spiders.

A wasp's sting is actually a modified egg-laying organ.

► WHY DO LEAFCUTTER ANTS CUT LEAVES?

Leafcutter ants are 'farmers'. They use the leaves they cut up to grow a crop of fungus, which supplies them with food.

Leafcutter ants can cut and carry pieces of leaf many times their own size. They are tropical ants that live in huge colonies. Workers cut up hundreds of plant leaves and carry the pieces back to the nest.

Inside the nest smaller workers chew up the leaves

into a kind of compost. They use this in special chambers to grow a special kind of fungus. The ants feed on the swollen juicy branches that the fungus produces. The fungus does not grow anywhere else and the ants tend it with great care. They weed out other fungi that may swamp it. They feed it with their droppings and treat it with their saliva to kill off any bacteria.

◄ WHICH INSECTS DRINK THROUGH STRAWS?

Many insects have 'straws' for sucking up liquids. The longest 'straws' are owned by some moths and butterflies.

Insects that suck up liquids have mouthparts that are extended into long tubes. Sucking insects include aphids, which have a sharp tube for piercing plant stems and sucking out sap. Mosquitoes pierce animal skin to suck up blood. A housefly has a tube with a sponge-like organ on the end for sucking up liquid.

The best 'straws', or proboscis, are found in insects that drink nectar from flowers. The long proboscis of a butterfly is normally kept coiled up and is extended only when in use. The convolvulus hawk moth has a proboscis 14 centimetres long for reaching deep into certain kinds of flower. One South American moth has a proboscis 30 centimetres long.

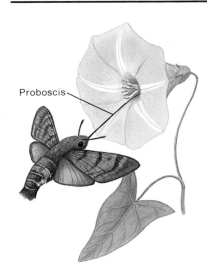

Proboscis

▲ WHICH FISHES KILL WITH ELECTRICITY?

Some fishes kill or stun their prey with electric shocks. Such fishes include the electric catfish, electric rays and the powerful electric eel.

The electric catfish (shown here) uses electrical discharges to stun and capture other fish and to warn off enemies. Its electric organ lies just under the skin and covers its body and part of the tail. A large catfish may produce 350 volts.

Electric rays are found in the warmer parts of the Atlantic Ocean. An electric ray has two separate electric organs, one on each side of the head. It catches a fish by pouncing on it, wrapping its fins around it and stunning it with electric shocks. Most electric rays produce about 50 volts.

The most powerful electric fish is the electric eel of South America. Its electric organ is in its tail and can produce 550 volts. This stuns or kills nearby fishes. Even a horse may be stunned by this eel.

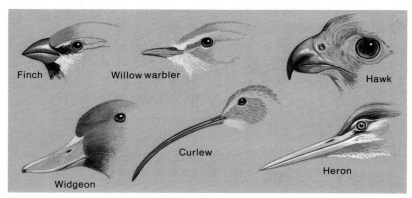

Finch Willow warbler Hawk Widgeon Curlew Heron

▲ HOW DO BIRDS FEED?

Birds may feed on insects, seeds, fishes, plants and small animals. You can often tell which food a bird prefers by looking at its beak.

Many birds that feed in similar ways have similar beaks. Birds of prey, such as hawks, have hooked beaks for tearing flesh. Seed-eating birds, such as finches, usually have short, conical beaks.

Some birds have very specialized beaks. The flat beak of a widgeon is used for cropping grass. A heron uses its dagger-like beak for spearing fish. Long beaks (for example, the curlew) are often used for probing, either in the ground for worms or in tree bark for insects. Willow warblers and swifts are insect-eaters.

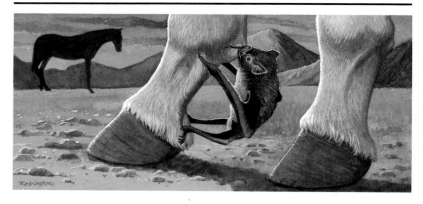

▲ DO VAMPIRES SUCK BLOOD?

Vampire bats are not the man-sized creatures seen in horror films. But these small mammals do drink the blood of other mammals, including humans.

Vampire bats are found only in Central and South America. They are small animals, no more than six to nine centimetres long.

Vampires feed at night. Common vampire bats suck the blood of mammals, such as horses and cattle. A vampire approaches its prey quietly, while the animal is asleep. With razor-sharp teeth it cuts away a small piece of skin, usually on the neck or leg of the animal. Its saliva contains a chemical that stops blood clotting. The vampire then sucks up the blood as it oozes from the wound.

▼ DO BEARS EAT HONEY?

Bears eat all kinds of things and most bears will eat honey when they can find it. The sun bear of southeast Asia is very fond of honey and is often called the honey bear.

Like Pooh Bear, nearly all bears are partial to a meal of honey. A bear will raid a nest of wild bees, tearing open the nest with its claws and licking out the honey with its long tongue. A sun bear will tear open a tree trunk to get at a nest inside.

Although bears belong to

the mammal order Carnivora (which means 'flesh-eaters'), they are actually omnivorous, that is, they eat all kinds of food. They like berries and other fruits and catch many kinds of small animals such as rodents, lizards, birds and insects.

A bear will even stand in a stream and catch fish, either by flipping them out on to the bank or by seizing them in the water. Some bears are experts at this.

▲ WHY DOES THE ELEPHANT HAVE A TRUNK?

An elephant's trunk is actually a combination of its nose and upper lip. It is used for many purposes, including feeding and drinking. Elephants acquired this useful 'tool' about 25 million years ago through evolution.

The ancestors of elephants were like large pigs. Gradually, they grew larger, but they had to develop long, sturdy legs. They also developed tusks and large, heavy heads. A heavy head cannot be carried on a long neck, so elephants need long trunks to reach their food.

An elephant uses its long, snake-like trunk for carrying food and water to its mouth and for spraying its body with water or dust. The trunk can be used very delicately – an elephant can pick up an object as small as a pea. It is also an efficient scent-detector, and elephants can often be seen with trunks raised, sniffing the scents in the air.

▲ WHY DOES A GIRAFFE HAVE A LONG NECK?

A giraffe's long neck and legs enable it to eat the leaves from branches that are far above the reach of other browsing animals. It can also see predators easily.

The giraffe is the world's tallest animal. An old male may reach a height of about five and a half metres. Being able to browse high in the trees and bushes is a great advantage to a giraffe. It has no competition from other browsing animals. And, as its eyes are so far above the ground, it can easily see predators long before they can get close enough to attack.

The disadvantage of long legs and a long neck is that a giraffe has to stretch a long way down to drink. However, it manages this quite well by splaying its front legs apart. A system of blood reservoirs and valves in its arteries prevents the giraffe from suffering a rush of blood to the head.

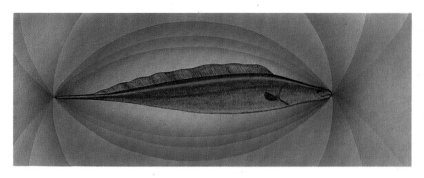

▲ WHICH FISHES 'SEE' WITH ELECTRICITY?

Some fishes that live in tropical rivers have little use for eyes because the water is too murky.

Instead they send out electric pulses and 'see' by detecting how the pulses are altered.

Fishes that navigate using electricity include the knife fishes of Africa, Asia and South America and the Nile fish of Africa. A Nile fish has a weak electric organ on its tail. This sends out a stream of electric pulses that form an electric field around the body (shown here). The pulses are picked up by special sense organs in the fish's head. Another fish or any obstruction nearby alters the pulses and the sense organs detect the change.

With this system of electric pulses a Nile fish can 'see' behind as well as it can to the front.

▶ WHY DO SOME ANIMALS HAVE NO EYES?

Animals without eyes live in dark places, such as caves. There, because there is no light to see by, sight is useless.

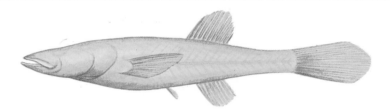

There are several species of blind cave fishes around the world. Their young often begin life with normal eyes, which then become useless as the fishes grow. In some species the remains of the eyes become covered with skin. Cave fishes find their way about by detecting pressure changes in the water around them. All fishes can do this, but in cave fishes the detection system is very highly developed. Some cave fishes explore their surroundings using taste buds on their lips.

Other blind cave dwellers include a number of beetles and the olm, a blind salamander. A young olm is black and has well-developed eyes. But as it grows older, its eyes become useless and it loses its pigment, becoming pink. Olms probably find prey animals by sensing the vibrations they make.

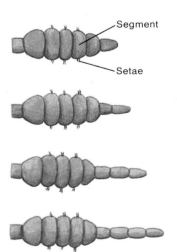

Segment

Setae

◀ HOW DOES A WORM HOLD ON?

An earthworm has no legs. But it does have tiny bristles that dig into the soil and help the worm to burrow.

On each segment of an earthworm's body there are four pairs of small bristles, or setae. When a segment shortens, it also expands and the setae are pushed outwards. When the segment lengthens, the setae are withdrawn. The setae of several segments together give the worm a firm grip.

The setae are mainly used in burrowing through the soil. An earthworm moves forward by first shortening and expanding the segments near its head. Then these segments are lengthened, one by one, and segments farther back are shortened. So a wave of 'shortening' seems to travel down the worm. As each segment shortens, its setae grip the sides of the burrow, allowing the lengthened segments in front to push forward.

▶ WHY DOES A SNAKE'S TONGUE FLICKER IN AND OUT?

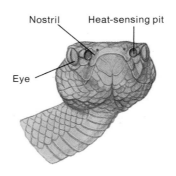

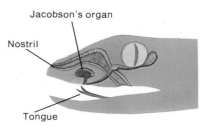

A snake 'tastes' the air by using a special organ in its mouth. It uses its flickering tongue to carry a stream of air samples back to this organ.

A snake's sense of smell is helped by a pair of special organs called Jacobson's organs. These are sensory sacs that lie side by side and open into the roof of the mouth.

When the snake's forked tongue flicks out of the mouth, it picks up chemical particles from the air. Inside the mouth the tips of the tongue are pushed into the Jacobson's organs which

'taste' the chemicals.

Some snakes, such as pit vipers and rattlesnakes, can also detect heat. They have special heat-sensing pits between their eyes and nostrils.

◀ HOW HIGH CAN A FLEA JUMP?

Claw   Suction pads

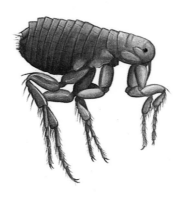

▲ HOW DOES A FLY WALK UPSIDE DOWN?

A fly can walk easily up a wall or across a ceiling because it has claws and suction pads on its feet.

The legs of all arthropods (the group that includes insects, spiders and crabs) are made up of several jointed segments. On the last segment of each leg a housefly has a pair of claws and two suction pads. These enable the fly to walk across almost any surface, even upside down.

For flying a housefly has a single pair of wings. Instead of hindwings it has a pair of club-like organs, called halteres. These vibrate rapidly when the insect is flying and act like gyroscopes to keep it stable. They also enable the fly to change direction suddenly. So it can make sharp, right-angle turns in mid-flight. And they help in landing. A fly approaches a ceiling flying in a normal, upright position. At the last moment, it performs a rapid backward half-somersault and lands upside down.

Fleas are tiny parasites with incredibly powerful hind legs. A flea can jump over a hundred times its own height. This is very useful when transferring from one animal host to another.

Fleas live in the fur or hair of animals. Their flattened bodies allow them to move between the hairs. They have no wings. Instead they have large hind legs for jumping.

A flea is less than one and a half millimetres long. But in one leap it can travel a distance of over 300 millimetres. And it can jump to a height of 190 millimetres. This is about the same as a human jumping to the top of a 65-storey office block.

To achieve this feat, a flea has very powerful hind leg muscles and tendons. In addition, the jumping apparatus of the flea has a piece of rubber-like material called resilin. When this is compressed and suddenly released, it produces a tremendous thrust.

▶ WHY DO SOME FISHES CLEAN OTHERS?

Cleaner fishes and their clients have an arrangement that is good for both of them. The cleaner fish gets food and the clients lose their parasites and other unwanted material. This partnership, where both sides benefit, is called a symbiosis.

A cleaner fish feeds on the tiny parasites that infest other fishes. However, it is quite a small fish and its clients may be large carnivores. To make sure that it does not get eaten, it has distinctive dark stripes. And before starting to clean, the cleaner fish makes sure that it will be accepted by performing a special kind of 'dance' in front of the client.

Often the client makes movements that show which parts of its body need cleaning. The cleaner fish then approaches and picks off parasites, dead skin and other debris from outside the clients body and inside its mouth and gills.

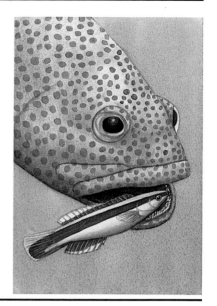

▲ WHY DO ANIMALS HIBERNATE?

Hibernation is a kind of deep winter 'sleep'. Animals hibernate to survive this cold period, when there is little or no food for them to eat.

Many animals hibernate during the winter. Some insects hide themselves away in nooks and crannies. Snails 'sleep' tucked up in their shells. Snakes, lizards, frogs and salamanders bury themselves in soil or mud.

A few birds, such as the poorwills of North America, hibernate. And a number of mammals, such as bears, hedgehogs, dormice and bats, go into a deep winter 'sleep'.

Hibernation begins as the days get shorter and the temperature drops. Mammals eat a lot of food before hibernating. They store it as fat, which their body processes use up very slowly while they are asleep.

▶ WHY DO A FISH AND A SHRIMP SHARE A HOME?

Some animals benefit each other by sharing a burrow or a nest. In the case of the *Alpheus* shrimp and the stargazer goby, the shrimp digs the burrow and the goby acts as sentry.

Alpheus shrimps dig burrows in the sand. There is usually a pair of shrimps in each burrow. They also share their home with a stargazer goby.

During the night both shrimps and goby rest in the burrow. During the day the goby places itself just outside the burrow, keeping watch for predators. The shrimps only feed when the goby is on sentry duty. As a shrimp feeds, it keeps one antenna touching the fish's tail. At the first sign of trouble, the fish flicks its tail and both animals retreat.

The goby cannot dig, so it benefits by getting a home. The shrimp's senses are not as good as the goby's, so it benefits by getting an early warning system.

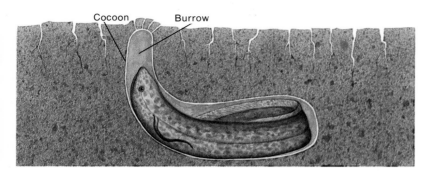

Cocoon   Burrow

itself a burrow with a hole or a plug of porous mud at the top. Inside the burrow it produces a large quantity of mucus. This hardens into a cocoon that stops the fish drying up.

A lungfish usually breathes air and it continues to get air through the hole at the top of the burrow. Its body processes slow down and it lives by absorbing some of its own muscle tissue. In this way it survives until the rains return and its stream fills up again. African lungfishes have survived embedded in dried mud for over four years.

▲ WHERE DOES A LUNGFISH GO WHEN ITS STREAM DRIES UP?

In dry weather a lungfish burrows into the mud. There it survives the

drought in a state of deep 'summer sleep' called aestivation.

As their streams dry up, some lungfishes tunnel into the mud. Each lungfish makes

▶ WHO FEEDS A CUCKOO?

Cuckoos are parasitic birds. They lay their eggs in the nests of other birds and let the other parents, or hosts, raise their young for them. The host parents' eggs and young are thrown out of the nest.

After mating, a female cuckoo keeps watch for small birds building their nests. She selects a nest in which one or more eggs has been laid. While the parents are away, she quickly flies down,

removes one egg and replaces it with one of her own. Common cuckoos lay their eggs in the nests of such birds as reed warblers, dunnocks and meadow pipits.

The cuckoo's egg is looked after by the host parents. But after it has hatched, the young chick pushes out the remaining eggs and any other chicks that have already hatched. The young cuckoo grows rapidly and soon becomes much larger than its foster parents. But they continue to feed the monster chick.

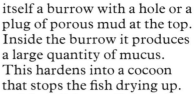

homes in the large swollen bulbs that develop at the bases of the thorns. To build their nests, the ants bore into the bulbs and dig out the material inside.

The ants defend the tree from caterpillars, aphids and all other plant-eating animals. They are rewarded for defending the plant. The tips of the leaves grow small sausage-shaped 'food-bodies', which are harvested and carried away by the ants. And the leaf stalks grow special nectaries, which provide the ants with a sugary drink.

▲ WHY DO SOME ANTS LIVE IN THORN BUSHES?

A whistling thorn has an armoury of sharp spikes to keep large, plant-eating animals away. But it also provides homes for fierce

ants, which keep off plant-eating insects.

A browsing animal that dares to try and eat the leaves of a whistling thorn is attacked by a horde of biting, stinging ants. The ants make their

25

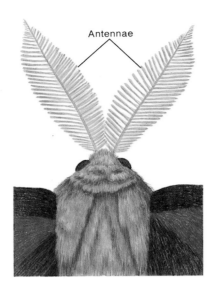

Antennae

▲ HOW WELL CAN A MOTH SMELL?

Moths often have large, feathery antennae. Using these, a male moth can detect the scent of a female several kilometres away.

Moths are mostly night-flying insects. So sight is not very useful in finding a mate. However, they have an amazing sense of smell. A male emperor moth can detect a female 11 kilometres away upwind.

Female moths produce special chemicals to attract males. They can even 'call' to males by inflating their scent glands and producing extra scent. The scent particles drift down-wind, spreading out in a plume.

The feather-like antennae of male moths have many highly sensitive scent receptors. And the antennae are spread out to catch as many scent particles as possible. Some moths, such as the silk moths, have massive antennae that can detect even the faintest trace of a female's scent.

▼ WHICH INSECT LAYS 1000 EGGS A DAY?

A queen termite is a huge egg-laying machine. All she can do is lay eggs, which she does at the rate of more than one a minute.

Termites live in large colonies. Some build huge, towering nests. At the centre of the nest lies the queen with her smaller king.

A mature queen has an enormously bloated body, up to 19 millimetres long, that serves as an egg-producing factory. At times she can produce over 30 eggs in a minute. She cannot move and has to be fed and tended by her workers.

A termite queen starts her life as a winged reproductive female. With many other winged females and males, she flies from the nest in which she was raised. When she lands, she mates with a winged male. Their wings drop off and together they found a new colony. The queen may live for over 20 years.

▲ WHY DO BEES SWARM?

A swarm of bees consists of a queen surrounded by a large number of workers and a few drones. They have left their old, over-crowded nest to found a new colony.

In late summer a colony of bees may contain over 60,000 workers and the nest may be overcrowded. At this stage new queens are produced together with a number of fertile males, or drones.

When the first new queen emerges, she usually kills the other queen larvae. Then she flies off, pursued by a number of drones. She mates with one of them and then returns.

If the existing queen is old, she dies when the new queen returns. On the other hand, if the existing queen is still young, she leaves her over-crowded nest, usually before the new queen hatches. She takes with her a swarm of workers and drones and they found a new colony.

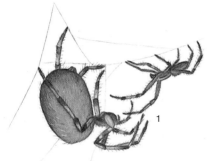

▲ WHICH MALES MUST BEWARE?

Some spiders and insects are so ferocious that the males must be careful how they approach their much larger females.

Black widow spiders get their name from the reputation of the females. They are said to eat their mates after mating. But this may not be entirely true.

Male diadem spiders are much smaller than the females, so they do have to approach with some caution (1). If they are careless they may be mistaken for prey. Usually, however, a male mates successfully, and lives to mate again. Just occasionally, when a male is exhausted after several matings, he makes the wrong signals and is devoured (2).

Male mantids, on the other hand, stand no chance. A male approaches a female very slowly and cautiously. At the last moment he jumps onto the female and mates with her. But within a few seconds she bites his head off and proceeds to eat him.

▶ WHY DO STICKLEBACKS CHANGE COLOUR?

A male stickleback puts on his 'wedding' colours to attract a female to his nest. He encourages her to lay her eggs, which he fertilizes and tends.

In spring, as a male stickleback begins to choose and defend a territory, his belly turns bright red. This is a warning to other males to keep out. He builds a nest of weed glued together in a shallow pit. Then his back turns bright blue to attract females.

When he has gained a female's attention, the male 'dances' under her belly and shows her the way into his nest. Once she is inside, he prods her tail to encourage her to lay eggs. After she leaves, he fertilizes the eggs.

The male tends the eggs. When they hatch, he guards the young until they can fend for themselves.

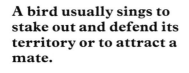

◀ WHY DO BIRDS SING?

A bird usually sings to stake out and defend its territory or to attract a mate.

Most birds begin to sing just before dawn. They sing to advertise their presence. In spring, many birds continue to sing all day. A yellow-hammer may repeat its 'little-bit-of-bread-and-no-cheese' over a thousand times before nightfall.

Birds can vary their songs. The song of a chaffinch has six possible variations. The chaffinch has 13 other calls. These may be used to tell of such things as the location of food or a good nesting site. Many birds sound an alarm. And different alarm calls can tell other birds whether the danger is in the air, in a tree or on the ground. Jays have several different alarm calls and can sometimes signal exactly which predator is nearby.

▶ WHY DO GREBES DANCE?

Grebes and many other birds perform ritual courtship displays before mating. It is a form of 'language' that birds use to attract mates of their own species.

Courtship displays have evolved over millions of years. They make sure that males and females of different bird species do not mate. If they did they would produce sterile eggs. Courtship is a special 'language'. A bird can indicate its species, its sex, where its territory is, when it is ready to mate and where there is a possible nesting site.

Grebes have several courtship dances. One of these is the 'head-shaking ceremony', in which the birds face each other and shake their heads from side to side. After head shaking, the birds may dive down to collect weed. Then they perform a 'penguin dance', in which they rise out of the water breast to breast and sway their weed-filled beaks from side to side.

▲ WHICH BIRD IS THE GREATEST WOOER?

A male bower bird goes to great lengths to attract a mate. He builds a bower, decorates it and entices a female into it.

The satin bower bird of Australia builds his bower during the summer. Using sticks, he constructs a platform and two thick 'walls' on the ground. Sometimes he uses a 'brush' made of fibres to paint the walls with the juice of berries or a mixture of saliva and charcoal.

At one end of the bower the male decorates a display ground with brightly coloured objects. These may be feathers, flowers, insect wings, berries and even man-made objects, such as tinfoil and bottle tops.

Eventually, a female comes to the bower and the male displays and calls to her. She enters the bower and, after she has rearranged some of the twigs, the birds mate. Then the female flies away and makes her own nest in a tree.

◀ WHICH BIRDS BUILD APARTMENT BLOCKS?

Weaver birds build elaborate, cosy nests. Many weavers live in colonies, and social weavers build huge communal nests in trees.

A weaver bird builds its nest high in the trees, usually suspended from a twig to help keep predators out. Generally, the male builds the nest. He weaves grass and strips of leaf into a hollow ball with a small entrance. The baya weaver of Malaya has a very neat nest with a long, downward-pointing tube as an entrance.

The social weavers of southern Africa build huge communal nests that may measure over four and a half metres across. They share the work of building a large 'thatched roof'. Then each pair builds a flask-shaped nest underneath. The 'apartment block' may be used again and again. Some colonies may be 100 years old.

► WHY DO MARSUPIALS HAVE POUCHES?

Marsupials, such as kangaroos and koalas, produce their young at a very early stage of development. The young crawl into their mothers' pouches, where they continue to develop in safety.

Most mammals nourish their developing young by means of a placenta. This is an organ that forms in the mother's womb and carries food and oxygen to the young. Most marsupials have no placenta and the young are born much earlier.

A young kangaroo is born 33 days after mating. It crawls, completely unaided, up through its mother's fur and into her pouch. There it fastens onto one of the nipples.

Two days later, the mother mates again. But the development of the second offspring is delayed. It is not born until seven months later. By this time the first offspring has left the pouch and only returns to it occasionally.

Kangaroos and koalas have pouches that open to the front. Burrowing marsupials, such as bandicoots, have pouches that open to the rear, so that soil does not enter as they dig.

► WHY DO BEAVERS BUILD DAMS?

A family of beavers build a lodge with underwater entrances to keep out predators. And to make sure that the water stays at the right level, they construct one or more dams.

A beaver lodge is built on an island of twigs, stones and mud just above the level of the beaver pond. Over this there is a dome of sticks and grass that is carefully plastered with mud. A small area at the top is left unplastered and this provides the chamber inside with ventilation.

A lodge built in a natural pond would flood during the rainy season and have its entrances exposed during the dry season. To avoid these problems a beaver family creates its own pond by building a dam downstream. Other dams may also be built upstream and below the main

dam. The dams are built using large, felled trees, boulders and small saplings. To seal a dam, beavers use sticks, leaves and mud.

Beavers work hard on their dams, heaving every tree and boulder into place. A dam rarely fails, partly because it is constantly being repaired.

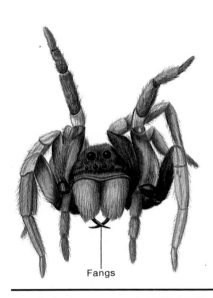

Fangs

◀ ARE ALL SPIDERS POISONOUS?

All spiders kill their prey by injecting them with poison. But only a few spiders are actually poisonous enough to cause large animals much harm.

A spider's jaws are equipped with a powerful pair of fangs. These are used to inject venom into its prey to kill it. So, as far as prey animals (mostly insects) are concerned, all spiders are poisonous.

However, few spiders have much effect on humans. A garden spider will sometimes bite. But such a bite is, at most, only mildly painful. Even one of the huge bird-eating spiders of South America has a bite no worse than a bee sting.

A few spiders are dangerous. One of the most notorious is the black widow spider. A bite from a European black widow can cause severe illness. And in North America black widow bites have proved fatal on a few occasions.

▲ WHY DOES AN ANTELOPE LEAP IN THE AIR?

Many antelopes, especially gazelles, leap high in the air. Sometimes they appear just to be playing. But they also leap to warn other antelopes that a predator is near.

The springbok is so called because of its ability to leap. When a grazing springbok sees a predator, it leaps suddenly into the air with its back arched and all four legs pointing straight downwards. The message is passed rapidly through the herd and soon all the animals are running and leaping away from the danger.

Impala, too, are well known for their extraordinary leaps. They may jump three and a half metres into the air. A series of leaps makes an impala look as though it is bouncing for joy. When disturbed by a predator, impala do not run in straight lines. They dodge about and leap as they go. A herd of animals moving in this way probably confuses predators.

▶ WHY DO BEES STING?

A bee stings to defend its nest. Any animal or human that interferes with a bee colony is liable to be set upon by a horde of worker bees.

Like the sting of a wasp, a bee's sting is its egg-laying organ shaped into a needle-like tube for injecting poison. A bee's sting is equipped with tiny barbs that prevent it from coming out. When the bee flies away, the sting is often ripped out. As a result the bee soon dies.

The European honeybee is fairly mild-mannered, although its sting is painful. However, there are other more ferocious bees. The giant honeybees of Asia are dangerous and people have been stung to death. In South America a strain of very aggressive African 'killer' bee accidentally introduced in 1957 is causing a problem. 'Killer' bees have chased out the local honeybees and are spreading rapidly.

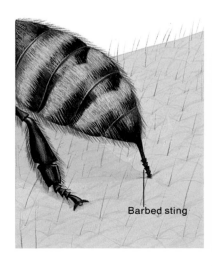

Barbed sting

▶ DO FLYING FISHES FLY?

Flying fishes travel long distances in the air. But despite their name, they glide rather than fly.

The most common flying fish is the two-winged type, which is found in all tropical seas. It is about 25 centimetres long and its pectoral (shoulder) fins are expanded into two large wings. The largest flying fish (45 centimetres long) is a four-winged type that lives off the coast of California.

A flying fish swims through the water until it reaches a speed of about 65 kilometres an hour. Just below the surface it starts to spread its wings and leaps out of the water. Then it glides through the air. Its momentum may carry it over 130 metres.

Flying fish leave the water partly to escape underwater predators. Unfortunately, when they are in the air, they are then prey to large seabirds.

◀ WHY DO HEDGEHOGS HAVE SPINES?

When frightened, a hedgehog curls up into a ball. Its spines give it excellent protection.

A hedgehog is born with a few rubbery prickles. Within seven weeks it has the thick coat of tough spines that will protect it for the rest of its life.

Hedgehogs mostly eat insects and other small animals. But a hedgehog will also tackle an adder. The adder finds it difficult to bite such a spiky creature. And in any case the hedgehog is immune to the adder's poison.

When a predator approaches, a hedgehog remains completely still until it has seen what the danger is. Then if really frightened, it rolls up with its head almost touching its tail. Only large carnivores, such as badgers and foxes, can penetrate this spiny ball.

Unfortunately, a hedgehog's defences also cause it a problem. It is hard to groom a skin covered in spines. As a result hedgehogs are infested with many small parasites.

▶ WHY ARE SOME SNAKES DANGEROUS?

Many snakes are deadly. Their fangs can inject poisonous venom.

Poisonous snakes can be divided into two groups, according to the position of their fangs. Snakes with poison fangs at the back of the mouth are not usually dangerous. They have some difficulty in sinking their fangs into a human.

The most dangerous snakes have fangs at the front of the mouth. These include puff adders, cobras, pit vipers, rattlesnakes and mambas.

Although many of these snakes have bites that can be fatal, most snakes do not attack humans unless provoked. Some give a warning before attacking; puff adders hiss loudly and a rattlesnake shakes its rattle. A few, however, such as the banded krait of Asia, the king cobra and the African mambas, may attack without any provocation or warning.

▲ HOW DOES AN ANIMAL BECOME INVISIBLE?

Many animals camouflage themselves to avoid being seen. The colouring and body patterns of a camouflaged animal enable it to merge with the background.

A number of animals are brightly coloured. But in their normal surroundings, this may be a great advantage. A bright-yellow crab spider remains unseen in a yellow

flower as it waits for its prey. Other animals, such as lions, need more drab camouflage colours.

Many animals camouflage themselves to avoid being eaten. Several species of moths rest on trees, where their markings blend with the bark and lichen. Other insects may be coloured green to match leaves.

Flatfish merge with the sand and pebbles on the seabed, but some fishes have irregular outlines as well as camouflage colours.

◄ WHEN IS A TWIG NOT A TWIG?

Some animals, especially insects, not only camouflage themselves by their colouring. They also disguise themselves as twigs or leaves.

Several kinds of insects look like twigs and stems. The picture shows a swallowtail moth caterpillar. Not only does its body have a bark-like colour and texture, but the insect also tries to act like a twig.

Stick insects are slow-moving animals. Their bodies are very long and they look like the stems of the plants on which they live. Different types have colours ranging from straw to brown and green. Some can even change colour to camouflage themselves.

Some green insects look like leaves. The leaf insect of southeast Asia is related to the stick insects, but its body is flattened and has 'leaf veins'. It looks exactly like a leaf, and even has slightly 'chewed' and battered edges.

▼ WHY ARE SOME ANIMALS BRIGHTLY COLOURED?

When animals are poisonous or unpleasant to eat, they sometimes advertise the fact with their bright colouring.

Bright warning colours warn predators that an animal should be avoided. A bird soon learns to leave the unpleasant-tasting cinnabar moth, with its red and black markings. Other animals with warning colours include the

Central American poison frogs, which are used to produce poison for arrow tips, and European wasps.

Some distasteful animals copy the warning colours of others. A cinnabar moth caterpillar is as unpleasant-tasting as the adult. To make sure its predators know this, it mimics a wasp's colouring.

Harmless animals also mimic the colours of poisonous ones. For example, a hoverfly looks very much like a wasp. The hoverfly gains protection from this colouring, because birds avoid it.

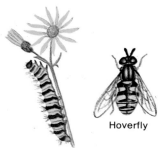

Hoverfly

Cinnabar moth caterpillar

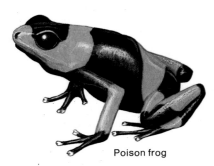

Poison frog

**Chameleons move slowly.
By changing colour to
match their surroundings,
they make it difficult for
their enemies and prey to
see them.**

The skin of a chameleon has
cells that contain coloured
pigments. Some cells contain
black pigment, others have
red or yellow pigment. By
using these pigments in
different ways, a chameleon
can change colour as it wishes,
sometimes very quickly.

A chameleon needs to be
camouflaged, or disguised,
because it may remain in one
place for a long time. When it
does move, it creeps very
slowly, so it is easy for
predators to catch it.

Chameleons feed by
remaining absolutely still
until a suitable prey comes
close enough. Then, faster
than the human eye can see,
the chameleon catapults out
its long, sticky tongue and
seizes its prey. Chameleons
usually eat insects and other
small animals.

▶ WHY DO SOME ANIMALS
HAVE FALSE EYES?

**One way of getting out of
trouble is to bluff. False
eyes can fool or startle a
would-be predator for
long enough to allow an
animal to escape.**

Several butterflies and moths
have false eyes on their hind
wings. When these are
exposed suddenly, an
approaching predator may be
startled into believing that it
is looking at a much larger
animal. By the time the
predator realizes its mistake,
the insect has flown away.
Other animals that have eye-
spots include caterpillars,
fishes and even a frog.

Some butterflies have not
only eye-spots but also false
antennae on their rear wings.
When the wings are in the

upright, closed position, the
back end of the insect appears
to be its head. A predator
trying to grab the insect goes
for a point just in front of the
'head' end and misses when
the insect flies away in the
other direction.

▼ WHY DOES A SQUID
SQUIRT INK?

**When it is alarmed, a
squid squirts a cloud of
black ink into the water.
This acts as a sort of
'smokescreen', behind
which the squid can
escape.**

Just below its head, a squid
has a small funnel. The squid
can contract its muscles and
force a jet of water through
this funnel. And, at the same
time, it can release ink into

the jet of water.

Squid can often avoid being
seen by camouflaging themselves. They are masters of
colour change and can
produce a wide variety of
patterns on their bodies.
When this fails, however, they
may escape by using jet
propulsion. At the same time,
they leave a smokescreen of
ink behind them. In some
species, the ink forms the
shape of a squid and the
predator attacks the ink decoy
while the real squid escapes.

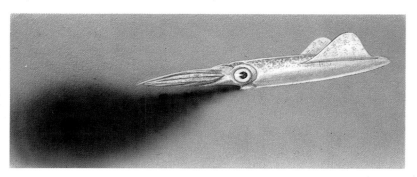

OURSELVES

▶ WHY DO THE POLICE USE FINGERPRINTS?

Whenever you touch something, your fingers leave behind marks, or fingerprints. Every person has his or her own unique set of fingerprints which are different from anyone else's. So fingerprints are used by the police to help to identify people who have committed crimes.

If you look at the tips of your fingers, you will see that each one has a pattern of tiny ridges on the surface. Your fingers also have sweat pores. So when you touch any object, you leave behind a slightly sticky impression of the ridges on your fingertips. These are called fingerprints.

The six most common kinds of fingerprint pattern are shown in the illustration. But these patterns can also have ovals, forks and other shapes in them. The result is that no two people have the same fingerprints. Even identical twins, who look exactly alike, have different sets of fingerprints.

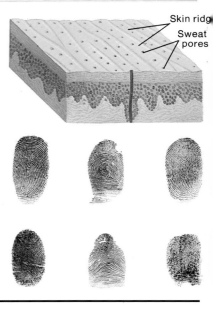

Skin ridge
Sweat pores

Suntanned skin

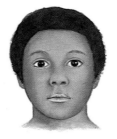
Black skin

Freckles

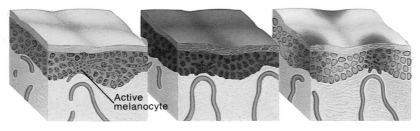
Active melanocyte

▲ WHAT DO A BLACK SKIN, A SUNTAN AND FRECKLES HAVE IN COMMON?

Skin colouring in all humans is controlled by the amount of the black pigment (colouring material) called melanin in the skin. A black person and a fair-skinned, suntanned person have different amounts of melanin. A freckled person has small spots of melanin in her skin.

The black pigment melanin is produced in special cells called melanocytes. All humans have about the same number of melanocytes in their skin. The darkness or lightness of the skin depends on the activity of the melanocytes. Dark-skinned people have very active melanocytes. They produce large numbers of melanin granules.

Melanin acts as a barrier against the harmful ultraviolet rays from the Sun. A fair-skinned person can increase the activity of her melanocytes by exposing her skin to sunlight. The skin darkens and she becomes suntanned. However, if she exposes untanned skin for too long, the melanocytes may not be able to produce melanin fast enough. Then the blood vessels swell and the skin becomes red and sore.

Some fair-skinned people have small groups of melanocytes that are more active than the rest. These produce dark spots, or freckles.

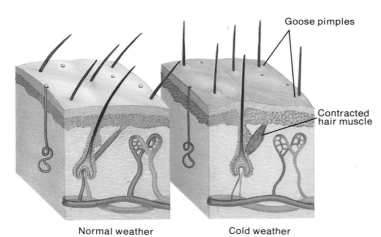

Goose pimples

Contracted hair muscle

Normal weather   Cold weather

When you are cold or frightened, the hairs on your body stand up. At the same time, the skin around each hair becomes raised into a 'pimple'. Your skin has a rough, bristly appearance, very much like the skin of a plucked goose.

All mammals are covered in hair. This insulates the body by trapping warm air next to the skin. In cold weather, the thickness of the layer of warm air is increased by raising the hairs. The erector muscle of each hair automatically contracts and the hair stands upright. This is why a cat looks larger on cold days.

Compared to other mammals, we humans have very little hair and so cannot trap much air next to our skin. We wear clothes instead. Even so, our hairs do stand up when we get cold. And as the erector muscle of each hair contracts, it draws together the skin around the base of the hair. The skin is pushed upwards to form a 'pimple'.

▲ WHY DO YOU SHIVER WHEN YOU ARE COLD?

When you are cold you need to warm up. Shivering is a form of movement that happens automatically and helps to generate heat.

When muscles contract and relax, they produce heat. This is why you get hot when running or digging in the garden. Sometimes, however, your body gets cold and your muscles 'switch on' automatically. They contract and relax rapidly, producing the movement we call shivering. In this way they generate some heat to help to overcome the effects of the cold.

Sometimes shivering is not enough to warm you up properly. This is why you often jump up and down and flap your arms on cold days. Again, the extra use of muscles produces more heat.

Exposure to cold conditions that chill the body can be very dangerous, especially to old people. They may suffer from hypothermia, which is a kind of drowsiness, followed by unconsciousness and sometimes death.

▶ WHY DO YOU SWEAT?

Sweating is a vital process for cooling you down when you get too hot. Sweat is produced by the sweat glands on the surface of the skin. There it evaporates and cools the body.

Your body prefers to operate at its normal temperature. So when your body temperature gets too high, such as during strenuous exercise, you need to lose heat. There are two ways in which this happens.

First, the tiny blood vessels in your skin increase in size and fill with blood, giving you a flushed appearance. Heat travels from the blood to the outside air.

When your body becomes even hotter, your sweat glands produce a mixture of water and waste chemicals known as sweat. When water evaporates, it uses up a great deal of heat (called latent heat of evaporation). So as the water in your sweat evaporates from your skin, it takes heat rapidly from your body.

▲ WHY DO SOME PEOPLE HAVE BLUE EYES AND OTHERS BROWN EYES?

You inherit the colour of your eyes from your parents. Special inherited factors, known as genes, **control the colour of your eyes and other features, such as your hair colour.**

Every person carries two genes for each feature. Often, one gene dominates the other. In the case of eye colour, a brown-eye gene dominates a blue-eye gene. If a person has two brown-eye genes then he or she will have brown eyes. Similarly, the presence of two blue-eye genes means blue eyes. But if a person has one brown-eye gene and one blue-eye gene, then the brown-eye gene dominates and the eyes are brown.

But a 'hidden' blue-eye gene is not lost forever, and can be passed on. In fact, it is even possible for two brown-eyed people, both with 'hidden' blue-eye genes, to produce blue-eyed children.

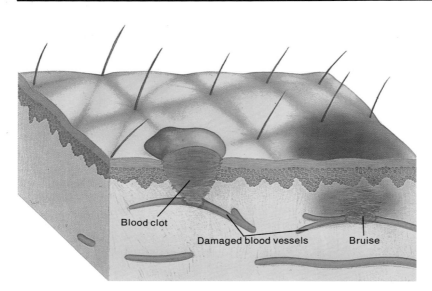

Blood clot

Damaged blood vessels

Bruise

▲ WHY DOES BLOOD CLOT?

When your skin is damaged, blood vessels near the surface are broken and you bleed. But unless the damage is severe, you do not bleed for long. Your blood contains special substances to stop the bleeding. These substances form a clot that dries into a hard scab.

Clotting of blood takes place very quickly. It begins when a substance called thrombin forms in the blood as it seeps from the wound. Thrombin is an enzyme. This is a chemical that causes a biological reaction to take place, without itself being used up in the reaction. Thrombin acts on a protein called fibrinogen and changes it to fibrin. This consists of many long fibres that become entangled with each other, trapping blood cells as they form.

Only the clear blood serum escapes, leaving behind a clot of fibrin and blood cells. This forms a protective covering over the wound.

◀ WHY DO BRUISES GO BLACK AND BLUE?

When an object strikes your body hard, the blood vessels beneath your skin may be damaged. If so, they release blood into your skin tissue, which becomes purple in colour.

The surface of your skin is relatively tough compared with the tissues beneath it. As a result, it is possible to damage these tissues without breaking the skin. Blood is released from damaged blood vessels and damaged cells also release fluid.

Often the area becomes swollen with excess fluid. When the damaged area is in a place where bone lies near the surface, such as the head or shin, the swelling has nowhere to go but outwards and a large bump appears.

The blood that enters the skin tissue shows as a purple discoloration. The blood cells break down and their contents are absorbed by the body again. While this is going on the colour of the bruise changes to brown and then yellow before it disappears.

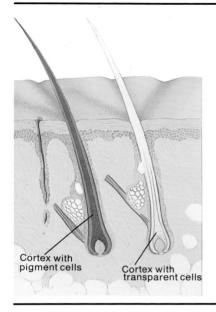

Cortex with pigment cells

Cortex with transparent cells

◄ WHY DOES HAIR GO GREY?

The colour of a person's hair is caused mainly by the presence of pigment-containing cells in each hair. In later life, some or all of the hairs grow without pigment and the colour appears grey.

The colour of a single hair is determined by cells in the hair follicle. This is the bulb-shaped structure at the hair root. These cells inject pigment granules (black, brown or yellow) into the cells of the hair cortex.

In early life, the colour of people's hair varies from black, through shades of brown and red, to fair and blonde. After a while, however, the pigment-producing cells in some hair follicles stop working. The hairs that grow from these follicles are actually colourless, but because of the refraction of light, they are seen as white. White hairs mixed with black or brown hairs give an overall grey colour to a person's hair.

▶ WHAT MAKES US CRY?

Our tear glands are constantly producing tears. Usually, they do not overflow, but gently bathe the front of our eyeballs before draining away. But sometimes, for example when we are upset, we produce tears faster than they can drain away. Then they flow down our cheeks.

The main purpose of the tears produced by our tear glands, or lacrimal glands, is to defend the eye against outside infection. They are mildly antiseptic and contain a substance that kills bacteria.

Tears are wiped over the eye regularly by blinking. This is an automatic action that occurs at least once every ten seconds. Excess tears drain away into the tear ducts and from there into the tear sac and the cavity of the nose.

Sometimes, because of pain or emotion, we produce a lot more tears than the tear ducts can cope with. When this happens, tears flow over our eyelids and down our cheeks. A blocked tear duct also causes tears to overflow.

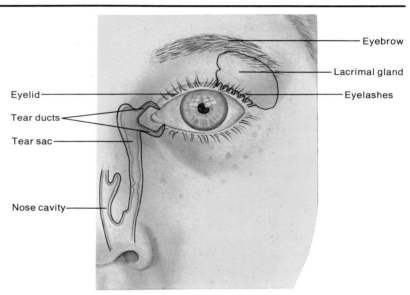

Eyebrow

Lacrimal gland

Eyelashes

Eyelid

Tear ducts

Tear sac

Nose cavity

▲ WHAT PROTECTS YOUR EYES?

Your eyes are very delicate organs that need protection. Tears kill invading germs and your eyelids, eyelashes and eyebrows help to keep out other harmful objects.

Your eyes' first line of defence are eyelashes and eyelids. Your eyelashes form two rows of stiff hairs around each eye. They help to catch and remove any large objects that come too close to your eyes.

Your eyelids are equipped with muscles so that they can close. The reflex action that makes you close your eyes helps to protect them from being injured by objects or dazzled by bright light. Any dust or dirt that does reach your eyes is removed when you blink. Your eyelids close briefly, sweeping across the front of your eyeballs.

The eyebrows form two long patches of protective hairs above your eyes. They prevent moisture from your forehead from running down into your eyes.

Foods containing iron

Foods for a balanced diet

Foods containing vitamin C

▲ WHY DO YOU NEED VITAMINS AND MINERALS?

Vitamins and minerals are present in your body in very tiny amounts. But they are very, very important to your health. Without them you would become ill.

You need only about one seven-hundredth of a gram of vitamin B_1 (thiamine) each day. But without it you would get a disease called beri-beri. Vitamin B_1 is found in bread and meat. All of the other 16 vitamins are also vital. For example, vitamin A (found in liver and carrots) is needed to help you to see in dim light. Vitamin C is especially plentiful in fresh fruit and vegetables. In olden times, lack of these foods on board ship led sailors to develop a disease called scurvy.

Minerals are also essential. You need iron (from meat, eggs and bread) in order to make the red blood pigment haemoglobin. Calcium (in milk, cheese and bread) and phosphorus (in most foods) are needed for the growth of bones and teeth.

▲ DOES IT MATTER WHAT YOU EAT?

In order to be healthy you must provide your body with the right raw materials. These are proteins, sugars, fats, vitamins and minerals. A balanced diet should include all of these.

You need energy to keep your body working. The energy value of food is measured in calories. The foods with most calories are those containing large amounts of carbo-hydrates (sugars and starches) and fats. Sugar, cereals, milk and cheese provide most of the carbohydrates and fats you need.

Proteins are also essential. An adult needs about 60 grams of protein every day to replace protein lost by wear and tear in the body. Growing children need protein to build new body tissue. Proteins are obtained from foods such as meat, fish, cheese and beans.

A balanced diet is completed by including fresh fruit and vegetables, which provide the remaining vitamins and minerals.

▲ DOES IT MATTER HOW MUCH YOU EAT?

If you do not eat enough food, you do not have enough energy to keep going. However, if you eat too much food, you do not use up all the energy it contains and it turns to unwanted fat.

The rate at which you use up energy depends on a number of things, including your age, weight and build. It also depends on what you do. For example, a man working in an office all day uses up fewer calories than a labourer on a building site. Active, growing children need a large number of calories. Old people need far fewer calories.

Ideally, therefore, everyone should eat food containing the right number of calories for their needs. Eating too much fat, carbohydrate and even protein leads to too many calories in the body. Excess calories turn to fat and so people who eat more than they need begin to become overweight. And people who are greatly overweight can suffer from poor health.

◀ WHY DO WE NEED EXERCISE?

Your body is a kind of machine. And like any other machine it needs looking after to keep it working properly. Exercise keeps your muscles working well. And general fitness helps you to be healthy.

An unfit person, who takes no exercise, converts only a small amount of food into energy and may become overweight. At the same time his muscles become weak and his blood circulation may become slow.

Exercise helps to make muscles stronger and improves their tone, or readiness for action. Well-toned muscles help to keep the bones properly placed in relation to each other. So a fit person has a better posture than an unfit person and is less likely to have backache.

Muscle movement helps to speed up blood circulation. At the same time exercise helps to increase a person's depth of breathing, making it easier to take in oxygen.

▶ WHAT IS YOUR BODY MADE OF?

Your body contains over 20 different chemical elements. The most plentiful element in the body is oxygen. Oxygen, together with hydrogen, forms water. Water makes up nearly two-thirds of your weight.

The body of an average person contains about 4.5 litres of water. It also has an amount of carbon equal to nearly 13 kilograms of coke.

Much of this carbon, together with hydrogen and oxygen, makes up fats and sugars. Carbon, hydrogen, oxygen and nitrogen form the body's vital proteins.

There are also large amounts of calcium and phosphorus. The body contains over one and a quarter kilograms of calcium and enough phosphorus to make over 2000 matches. The body also contains a couple of spoonfuls of sulphur, enough iron to make a 2.5-centimetre nail, and nearly 30 grams of other metals.

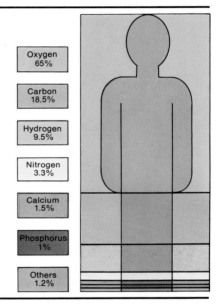

Oxygen 65%
Carbon 18.5%
Hydrogen 9.5%
Nitrogen 3.3%
Calcium 1.5%
Phosphorus 1%
Others 1.2%

▲ WHY DO YOU NEED REST?

You spend about one-third of your life sleeping. No one knows exactly why we sleep, but it seems to be essential for the health of both the mind and body.

The amount of sleep we need depends on several things, including a person's age and what he does during the day.

A new-born baby sleeps nearly all day. It only wakes up to be fed. A young child needs about 12 hours sleep a day. An adult generally needs to sleep for about eight hours, although some people stay healthy with only two hours sleep a night.

After a period of sleep you should wake feeling rested and refreshed. Someone who goes for more than 36 hours without sleep becomes irritable and confused. If he manages to stay awake without collapsing for over 60 hours, he may begin to 'see' and 'hear' things that are not there.

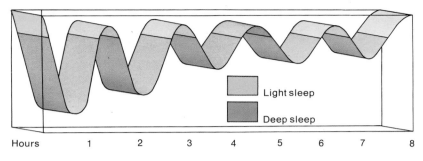

| Hours | 1 | 2 | 3 | 4 | 5 | 6 | 7 | 8 |

Light sleep

Deep sleep

▲ WHY DO YOU DREAM?

Sleep is an essential part of our lives (see previous page). It is a kind of unconsciousness, but we can wake fairly easily. Even though the conscious part of the mind is totally at rest, the subconscious part of the mind continues working and produces dreams.

When you fall asleep, you quickly go into a state of deep sleep. But about five times during the night you change from periods of deep sleep to periods of light sleep. And as morning approaches, the periods of light sleep become longer.

You dream during the periods of light sleep. At these times you make rapid eye movements and some body movements. If you wake up while you are dreaming, you can often remember the dream. No one really knows why we have dreams. But they appear to be an important part of sleep.

► WHY IS YOUR TEMPERATURE TAKEN WHEN YOU ARE ILL?

Your normal body temperature is about 37°C, or 98.4°F. An increase in temperature may be a sign of illness.

Body temperature is measured by using a clinical thermometer. This is a glass tube with a bulb containing mercury at the end. Expansion of the mercury in the bulb causes it to move up the tube. A scale on the side shows the temperature.

A person's temperature is usually taken by placing a thermometer in the mouth for about three minutes. It can also be taken by putting the thermometer in the armpit, but this may show a lower temperature.

Normal temperatures vary between 36°C and 37°C. A temperature above this is often a sign of infection. However, a high temperature is not always a sign of illness. On the other hand, a person may be very ill yet still have a normal temperature.

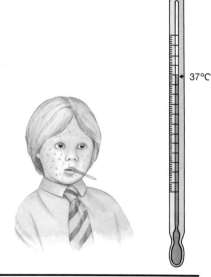

← 37°C

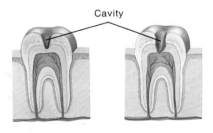

Cavity

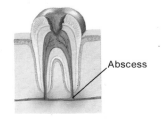

Abscess

◄ WHY SHOULD YOU BRUSH YOUR TEETH?

Teeth have hard surfaces that seem impossible to get through. But they can be attacked and eaten away by bacteria. These tiny living organisms thrive on any food that is left on or around your teeth. So to prevent your teeth from decaying, you should brush them regularly.

Food debris on a tooth begins to form a hard material called plaque. Bacteria grow in plaque and start to eat away the enamel layer of the tooth. If this decay is not stopped, you may eventually have a cavity that reaches into the inner pulp of the tooth. Then, because nerve endings are exposed, you will have a very painful toothache. If the bacteria infect the tooth right down to the root, they then attack the bone in which the tooth is set. This causes an extremely painful abscess.

To prevent this you should brush your teeth after every meal.

▶ WHY IS SMOKING BAD FOR YOU?

Cigarette-smoking is a major cause of death. By breathing in, or inhaling, tobacco smoke you greatly increase your chances of getting lung cancer. You may also suffer from heart disease and chronic bronchitis. Smoking is difficult to give up. So it is best not to start at all.

Tobacco smoke contains many different substances. Some of these are poisons, including nicotine, the drug that is thought to make smoking difficult to give up. Others have been shown to cause cancer in animals.

Some chemicals affect a smoker's air passages and lungs, causing bronchitis. Other chemicals are absorbed into the blood and affect the organs of the body, including the heart.

It is not known exactly how smoking causes disease. But it is certain that smokers die at a younger age than non-smokers (on average). The heaviest smokers run the greatest risk of dying early.

▲ IS ALCOHOL GOOD OR BAD?

In small amounts alcohol can calm nervous tension and stimulate the appetite. But in larger quantities alcohol can be dangerous. It reduces a person's ability to control his own actions and it also poisons the body.

A person who takes one or two drinks a day does no harm to his health. In fact, he may even benefit from the relaxing effect of the drinks.

However, greater quantities of alcohol begin to be harmful. Alcohol acts by numbing parts of the body's central nervous system. After several drinks a person may feel happy and carefree. But he is in fact unable to cope with tasks that require skill and judgement.

More alcohol causes an even greater loss of coordination and the drinker may become unconscious. The alcohol poisons his bloodstream, causing sickness and leaving him with an upset stomach and a hangover.

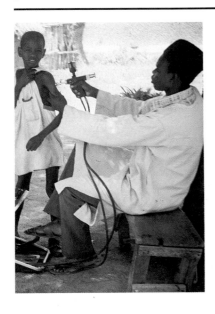

◀ WHY ARE VACCINATIONS NECESSARY?

Vaccination is an important method of controlling the spread of many infectious diseases. A person given a vaccine is made immune to the disease and cannot pass it on to anyone else.

Diseases that can be prevented by vaccination include polio, tetanus, whooping cough, diptheria, measles, German measles, cholera and typhoid. Small-pox was once a major disease, but has now been wiped out by vaccination.

Vaccines may be given by injection, by mouth or by a scratch on the skin. Vaccine causes the body to produce antibodies, which neutralize invading germs. Some vaccines contain dead germs. Others contain toxoids – chemically modified toxins (poisons) normally produced by the disease-causing germs. Other vaccines contain live germs that are very closely related to the disease-producing germs.

SCIENCE

▶ WHY CAN YOU SEE YOURSELF IN A MIRROR?

You can see yourself in a mirror because light rays bounce off its shiny surface. Light rays come from everything you can see, including yourself. You see things when the light rays from them enter your eyes. Some of the light rays that come from yourself strike the mirror. The mirror reflects the rays because it is very smooth. The rays come back to you and enter your eyes.

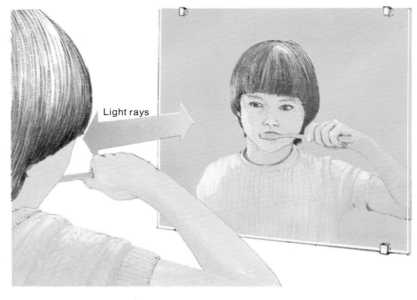

Light rays

The light rays that come from things move in straight lines. However, when they are reflected from a mirror, the light rays change direction but still travel in straight lines. As light rays are invisible, you cannot see that the rays from yourself have

been reflected by the mirror. Your brain is tricked into thinking that the rays have come direct to you in one straight line without being reflected, as if a copy of yourself were standing on the other side of the mirror. You therefore see an image of yourself in the mirror.

The image is back-to-front. This is because the mirror reflects light rays from one side of yourself (the left side, say) back to you on the left. In the mirror, your left side appears to be on the left. But when real people face you, their left side is always to your right.

Gravity pulls large objects towards each other

◀ WHY DO THINGS FALL TO THE GROUND?

Anything that is dropped falls to the ground because it is pulled down. An invisible force called gravity makes things fall. If there were no gravity, everything would float about.

The Earth is not the only thing to possess gravity. Everything has gravity in fact. You do, and so do the Sun and Moon. However, the amount of gravity things have

depends on how big they are. A massive object like the Earth has a strong force of gravity.

When two things are free to move, the gravity between them pulls them together. The Earth pulls at you and you pull at the Earth. But because the Earth's gravity is so much stronger than yours, you move towards the Earth. That is, you fall. On the Moon, gravity is six times weaker than the Earth's gravity. You would fall slowly on the Moon, but you could also jump much higher.

▶ WHY DOES GLUE STICK THINGS TOGETHER?

Glue contains chemicals. When glue sets, the chemicals change and become new chemicals that pull very strongly on things. When you stick something with glue, it pulls the two parts together so strongly that they become fixed to the glue and therefore to each other.

The chemicals in the glue are made of very small particles called molecules. All molecules pull on each other with a strong force. In fact, this force keeps all solid objects together in one piece. If two things are placed very close together, like two pieces of plastic film, the molecules in one surface pull at the molecules in the other and they stick together.

Glue acts in a similar way. It fills the tiny gaps between the surfaces of the two parts, so that the molecules of glue pull at the molecules in the surfaces. As the glue sets it sticks the parts together.

◀ WHY DOES SOAP MAKE BUBBLES?

If air is blown into water, bubbles rise to the surface. The water squeezes the air in the bubbles strongly and they burst at the surface. If the water is soapy, it does not squeeze the air so hard. A film of soapy water stays around the air, and bubbles float above the water.

The molecules in water pull at each other with a strong force. At the surface of the water, molecules are pulled by the other molecules in the surface and below. As there are no water molecules above the surface, there is more force between the surface molecules than between the molecules underwater, which are pulled on all sides. This extra force is called surface tension.

As a bubble breaks the surface, the surface tension normally pulls the water back and the bubble does not continue. But soap lowers surface tension so that a film of water can form around the air and continue the bubble.

Air inside bubble

Film of soapy water

▶ WHY DOES PAPER TEAR EASILY?

Paper may look smooth and solid, but it is not. If you could see it magnified, you would see that it is made of many tiny fibres all pushed together. If you pull on the paper, the fibres easily come apart and the paper tears.

Paper is made by beating wood and rags with water to make a pulp, and then spreading out the pulp into a thin layer and drying it. A certain amount of glue may be added to the pulp to help stick the paper fibres together. Little glue is added when making newspapers or paper tissues. The paper fibres are held loosely together and this kind of paper tears easily.

The kind of paper that is used to make book pages contains more glue and it does not tear so easily. Cardboard is a thick kind of paper containing strong glue, and it is difficult to tear it. Soaking paper in water may dissolve the glue so that it then tears very easily indeed.

43

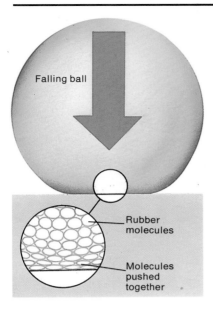

Falling ball

Rubber molecules

Molecules pushed together

▲ WHY DO RUBBER BALLS BOUNCE?

If you pull a rubber band, it stretches. When you let go, the band immediately springs back to its original size. A rubber ball is like a band in reverse. When the ball hits the ground, it gets smaller. Then it springs back to its original size. As the ball gets bigger, it pushes on the ground. As it does so, it pushes itself up into the air.

Rubber bounces because it is elastic. When it is stretched or squeezed and then released, it immediately springs back to its original shape. It does this because of the force between its molecules.

The molecules like to be at a certain distance from one another. If they are pulled apart, which happens when the rubber is stretched, the force between the molecules acts to pull them back together. Similarly, if they are pushed together, which happens when the rubber is compressed, the force between the molecules acts to push them apart again.

▼ WHY DO THINGS GO RUSTY?

Iron objects go rusty after a time. Steel goes rusty too, because it contains iron. The rust is caused by oxygen (an invisible gas in the air) and by water. The water may also be in the air in the form of water vapour. The oxygen and water together turn iron into rust.

The oxygen in the air attacks iron to form a chemical compound called iron oxide. It needs water to do this, which is why iron and steel objects quickly go rusty in wet places. Iron oxide is a reddish-brown colour, so the iron or steel develops a reddish-brown covering as it begins to rust. This continues until the whole object changes into rust. The rust is not strong like iron and steel, so rusting makes things weak.

Rusting can be prevented by covering the metal with paint. The layer of paint stops the oxygen from reaching the iron or steel beneath. Some kinds of steel contain other metals that stop them rusting.

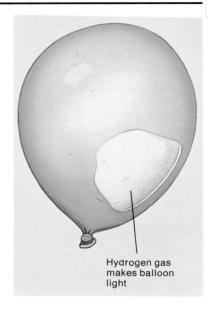

Hydrogen gas makes balloon light

▲ WHY DOES A BALLOON FLOAT IN THE AIR?

If a balloon is blown up with a gas called hydrogen instead of air, it floats up into the sky. This is because hydrogen is lighter than air. It rises in air just as a bubble of air rises in water. The hydrogen is light enough to carry the weight of the balloon.

The hydrogen in the balloon does have some weight. However, hydrogen has a lower density than air. This means that the balloon weighs less if it is filled with hydrogen than if it were filled with air. As the balloon is lighter than the air, it rises up into the air.

If the balloon is blown up with air, the air inside is not lighter or heavier than the air outside. But the rubber of the balloon makes it heavier than the air, and it sinks instead of rising.

Hydrogen is made up of particles called molecules. It is a very light gas because its molecules are smaller than any other kind of molecules.

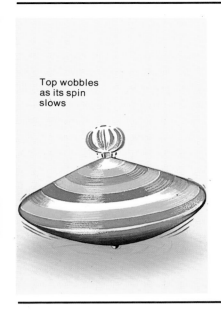

Top wobbles as its spin slows

◀ WHY DOES A SPINNING TOP NOT FALL OVER?

If you spin a top, it stays upright as long as it spins quickly. Then it begins to wobble as it slows down. Finally, it tilts over so much that it hits the ground and stops. When anything spins, it cannot be pulled over. The force of the spin stops the downward movement and makes the spinning object tilt sideways. The tilt gets bigger as it slows down, and the top wobbles.

When the top is exactly upright, it balances perfectly. However, it then begins to fall over and the upper part of the top moves slightly downwards. As it does so, the spin of the top changes the downward movement into a sideways movement. It makes the upper part of the top move in a circle and the top tilts over as it spins. The amount of sideways movement depends on the speed of spin. As the top slows down, it gets less, allowing the downward motion and the tilt to increase.

▶ WHY DO WET THINGS SHINE?

Anything that is completely smooth looks shiny. This is because it reflects all the light towards you. A rough surface does not do this if it is dry. But if it is wet, the water makes it smooth because it fills in the tiny holes in the surface. It therefore looks shiny.

A smooth surface reflects all the light rays that strike it at the same angle. This means

Light rays

Light reflected by smooth surface

Light spread by rough surface

Thin film of water

that more light may be reflected by the surface in one direction than in another. This is why light seems to shine more brightly from one part of a shiny surface than another. The surface of water is completely smooth and

therefore is shiny. A wet surface therefore shines as well.

Light rays striking a rough surface are reflected at different angles by different parts of the surface; so it does not look shiny.

▶ WHY ARE THERE ONLY TEN NUMBERS WITH SINGLE FIGURES?

All decimal numbers are made from ten figures. These are 0, 1, 2, 3, 4, 5, 6, 7, 8 and 9. There are only ten of these figures because we probably began counting on our fingers and thumbs.

We could have any number of single figures other than ten. However, using ten makes it easier for us to learn how to handle numbers, because we

can use our fingers to count.

Roman numerals have letters to stand for numbers. V is 5 and C is 100, for example. This system makes calculations like multiplying and dividing very difficult to do. Having a figure for zero, and several other figures from which all numbers can be made, makes calculations much easier.

Computers use the binary system, which has two figures: 0 and 1. To make calculations, a computer converts decimal numbers into binary ones.

Roman numerals

I II III IV V VI
VII VIII IX X

Decimal numerals

123456
7890

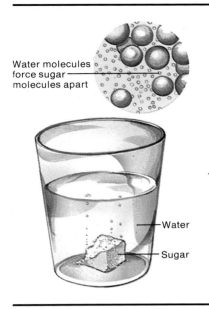

Water molecules force sugar molecules apart

Water

Sugar

◀ WHY DOES SUGAR DISSOLVE IN WATER?

If you put some sugar in water and stir it, it disappears. However, the water tastes of sugar. The sugar dissolves, or enters the water. This is because it breaks up into tiny particles that spread out through the water.

Inside sugar crystals, molecules of sugar are lined up neatly in rows. As the sugar is placed in the water, water molecules invade the crystals. The water molecules come between the sugar molecules and force them apart. The sugar molecules move out among the water molecules as the sugar dissolves in the water. Heating the water makes the sugar dissolve quickly because it makes the molecules move faster.

In many substances, the forces that hold their molecules together are too strong to allow water molecules to force them apart. These substances therefore do not dissolve in water.

▶ WHY DOES A BOAT FLOAT ON WATER?

A boat can float on water even if it is made of metal, which is heavier than water. This is because the boat pushes aside some of the water. The water pushes back on the boat, and supports its weight. This makes it float.

The support that a boat gets from the water is called up-thrust. The amount of up-thrust is equal to the weight of the water that the boat pushes aside or displaces. If the upthrust produced is equal to the weight of the boat, then it will float. The boat must therefore displace a large amount of water in order to float. This is why boats are hollow and broad in shape.

If the boat were made of solid metal or was narrow so that it was much smaller in size but the same weight, it would not displace as much water. The amount of upthrust produced would be not enough to support the weight of the boat and it would sink.

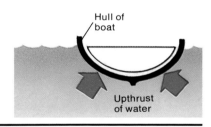

Hull of boat

Upthrust of water

◀ WHY DOES WATER FLOW MORE EASILY THAN SYRUP?

Syrup flows very slowly from an upturned tin or jar because it sticks to the sides of the container. Water does not stick as much as syrup, so it flows more easily.

The molecules of syrup pull on each other with a strong force. They also pull strongly at the molecules in the sides of the container. These forces or bonds between the molecules act to slow down the movement of the syrup.

The bonds between water molecules are not as strong as the bonds between syrup molecules. With water, movement is not slowed as much as with syrup, so water flows more easily.

Syrup flows faster if it is warmed. This happens because the heat causes the syrup molecules to move about within the syrup more quickly. The extra movement weakens the bonds between the molecules in the syrup, and as a result the warm syrup flows faster than cold syrup.

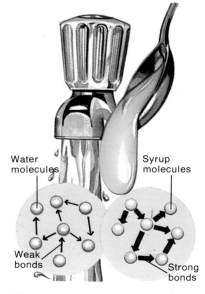

Water molecules

Syrup molecules

Weak bonds

Strong bonds

▶ WHY DOES WATER FREEZE IN WINTER?

It may get so cold in winter that the tempera-ture falls to freezing point (0°C) or lower. When water gets this cold, it freezes.

Water is made up of molecules that move about. This is why water is liquid and flows easily. The speed of movement depends on the temperature of the water. The hotter it is, the faster the molecules move. As the temperature falls, the molecules gradually get slower and slower.

At freezing point, their movement becomes so slow that each molecule begins to exert a pulling force on the molecules around it, and they begin to line up in rows. The molecules do not stop moving, but each one now vibrates around one position instead of moving about freely. As the molecules take up their positions, the water changes from being a liquid and becomes a solid. That is, it turns into ice.

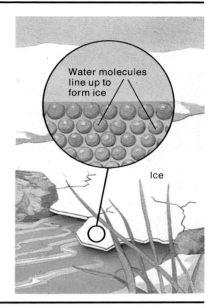

Water molecules line up to form ice

Ice

◀ WHY DOES A PUDDLE DRY UP IN SUNSHINE?

The warmth of the sunshine makes some of the water turn into water vapour. Water vapour is an invisible gas which mingles with the air. Slowly, all the water turns into water vapour, and the puddle dries up.

When water turns into water vapour, it is said to evaporate. This happens because the molecules of water are constantly moving within the water. At the surface, some molecules escape from the water and enter the air. In this way, water vapour is formed. Evaporation continues in this way if a wind is blowing to carry away the water vapour, or if it is warm and dry.

Usually, the molecules continue to escape and the water turns completely to water vapour. However, the air may become so laden with water vapour that no more water molecules can escape from the surface. The air becomes very humid, and puddles do not dry up.

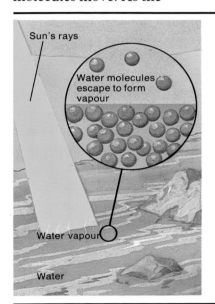

Sun's rays

Water molecules escape to form vapour

Water vapour

Water

▶ WHY IS ICE SLIPPERY?

If you try to pick up a piece of ice, it usually slips out of your fingers. This is because the warmth of your fingers melts the surface of the ice and turns it into water. A film of water therefore forms between your fingers and the ice, making the ice feel slippery.

If the ice is very cold, it feels sticky instead of slippery. This is because the warmth of your fingers first melts the surface of the ice, but the ice is so cold that this water im-mediately freezes again, making your fingers stick to the ice.

Ice skaters are able to skate on ice for a different reason. The weight of the skater's body throws great pressure on the ice as the skates pass over the ice. This pressure causes the ice to melt beneath the skates, so that the skater in fact slides on a film of water between the skates and the ice. This film immediately freezes to ice again after the skater has passed.

Skate

Film of water

Ice

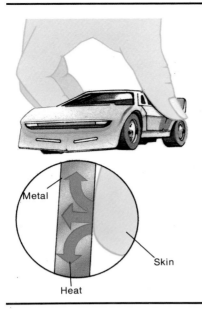

Metal

Skin

Heat

◄ WHY DOES METAL FEEL COLD?

When you touch something made of metal, it feels cold. The reason for this is that some heat flows from your fingers into the surface of the metal. Because the skin on your fingers loses heat, it goes cold and the metal feels cold.

Metal objects feel cold in a cool climate because metal is a good conductor of heat. This means that heat flows through metal easily. Heat therefore leaves the skin of your fingers and flows into the metal. It does this because your fingers are warmer than the metal. The heat moves on through the metal, so that the surface of the metal does not get as warm as the skin on your fingers. Heat continues to flow from your fingers into the metal, and it feels cold.

Wood or cloth are poor conductors of heat. The surface of an object made of wood or cloth quickly warms up as you touch it, and it does not feel cold.

► WHY DOES WOOD BURN BUT NOT IRON?

For anything to burn, it needs the oxygen gas that is in the air. When wood burns, it takes in oxygen. Together, the wood and oxygen give out heat and form ash. The wood has to be heated to make it take in oxygen. When it is hot enough, it catches fire and begins to burn. Iron does not take in oxygen like this when it is heated. It therefore does not catch fire and burn.

Things like wood burn because the oxygen molecules in the air split apart the molecules in the wood, forming molecules of ash and also gases, including carbon dioxide. To do this, the wood and oxygen molecules need a certain amount of energy so that they will move or vibrate faster and overcome the forces that keep them apart. This energy comes from heat.

Iron molecules do not split as wood molecules do, no matter how hot the iron is heated.

Iron does not combine with oxygen and burn

Wood combines with oxygen and burns

◄ WHY DO CLOTHES KEEP YOU WARM?

Clothes may feel cold when you first put them on. But they soon warm up and then keep you warm throughout the day. This is because your body produces heat. The clothes stop most of this heat from leaving your body.

Clothes are good insulators. This means that they do not conduct heat very well. The heat of your skin first warms the clothes, and then little heat flows from your skin into them. Some heat does get through, particularly if the outer surface of the garments is cold. This is why you put on more clothes or thicker garments to keep warm on a cold day.

The main reason why clothes are good insulators is that they contain air. Air is trapped between the fibres in the cloth. Also, a layer of air is caught between the clothes and your skin. Air is a poor conductor of heat, and so clothes keep your body heat in.

Cold air

Cloth

Layer of air

Warm skin

▼ WHY IS A FLAME HOT?

When a candle or a fire is lit, it burns with a hot flame. It has to be heated with a match or a lighter to catch fire. As the burning continues, the candle or the fuel in the fire takes up oxygen in the air. As it does so, it gives out heat and the flame continues to be hot.

In anything that is hot, the molecules are moving or vibrating quickly. The hotter it is, the faster the molecules

Oxygen

move or vibrate. In a flame, oxygen in the air is combining with a fuel, for example wood.

The molecules of wood and of oxygen each break apart and then come together to form new molecules of ash and also of invisible gases like carbon dioxide, which escape into the air. As they do so, energy is given out by the old molecules and taken up by the new molecules. However, the new molecules do not take up as much energy as that given out. The extra energy makes them move faster and causes heat to be produced as the wood burns.

▼ WHY DO MATCHES LIGHT WHEN YOU STRIKE THEM?

When you strike a match, you rub the head of the match against the rough surface on the side of the matchbox. This makes the head of the match get hot. The head contains substances that burst into flame when they get hot. This is why the match lights.

As the head of the match moves over the rough surface, the molecules in the head and

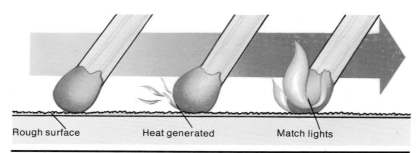

Rough surface   Heat generated   Match lights

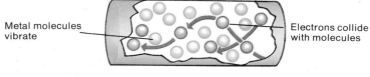

Metal molecules vibrate

Electrons collide with molecules

▲ WHY DOES AN ELECTRIC FIRE GLOW RED?

When you switch on an electric fire, electricity flows through the wire in the bars. This makes the bars get hot. If anything gets hot enough, it gives out light. First it glows red, then yellow and finally white if it gets really hot. Electric fires get hot enough to glow red.

An electric current consists of a flow of tiny particles called electrons. As the electrons

the surface collide with each other and move faster. This makes the head of the match hotter. In the head are substances that catch fire without much heating. The heat produced by striking the match is enough to make the head burst into flame.

In safety matches, the side of the matchbox contains a substance that is needed to make the head of the match catch fire. These matches therefore do not light when struck on any other surface.

pass through the wire in the bars of an electric fire, they collide with the molecules of metal in the wire. This makes the molecules vibrate faster and the metal gets hotter. In this way, the energy of the electricity is changed into heat energy.

At high temperatures, things give out light energy as well as heat energy. At very high temperatures, the light has a lot of energy and is white. But at the lower temperature of an electric fire, the light has less energy and the bars of the fire glow red.

TRANSPORT

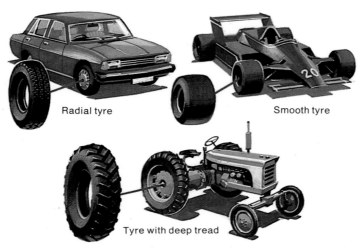

Radial tyre

Smooth tyre

Tyre with deep tread

Most modern cars have radial tyres, so called because their internal cords run radially. These are suitable for use on most road surfaces and in most weather conditions.

In snowy areas, however, motorists have to fit winter tyres. These have deeper treads and often metal chains as well to obtain a greater grip on slippery surfaces.

Special-purpose vehicles have tyres designed to suit specific conditions. Completely smooth tyres permit very high speeds without loss of grip in dry weather and are used on racing cars. Tractors and earth-moving vehicles have tyres with huge treads to help them travel over rough ground.

▲ WHY ARE THERE DIFFERENT TYPES OF TYRE?

Tyres are the point of contact between a vehicle and the road. They grip **the road, enabling drivers to brake, stop and steer their vehicles. They also absorb some of the vibrations from the surface of the road.**

▼ WHY DO CAR DESIGNS DIFFER?

A car's design has an important effect on its performance, the speed at which it travels and the amount of fuel it uses.

Fashion also influences the styling of new models.

The first cars were called horseless carriages. And they were just that: a vehicle built by a traditional coach-builder but with a motor instead of a horse. Each chassis was built by hand, and there was a large variety of styles.

In the 1920s, when factory mass production was introduced, styles became more standard. The classic car of this period is a sedate box on wheels. In the 1950s and 1960s, especially in the USA, the fashion was for large 'gas-guzzlers' with huge fins and bumpers. Since 1973, when oil prices started to rise, smaller, more economical cars have been in fashion.

The more streamlined a car, the more smoothly air flows round it as it moves along, and the less fuel it will use. The trend is for car designs to become even more streamlined.

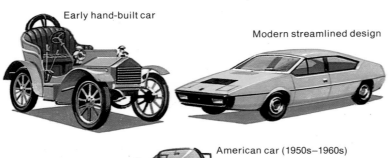

Early hand-built car

Modern streamlined design

American car (1950s–1960s)

▼ WHY DO WE WEAR SEAT BELTS?

Seat belts are a very important safety device. If your car stops suddenly, they hold you in your seat and help to prevent serious injuries.

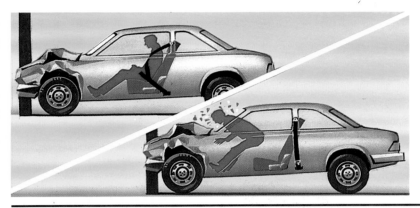

If two cars travelling at 60 kilometres per hour hit each other, the force of the crash is 120 kilometres per hour. If you are wearing a safety belt, with luck you will probably just be jerked sharply forwards. If you have not fastened your seat belt, you will be hurled through the windscreen. At the least, your face could be badly cut. But you could be thrown on to the road outside with bad injuries and you might be killed. If you are driving, you may find yourself crashing into the steering wheel.

Seat belts save lives. In many countries, front-seat passengers must wear them. Belts for rear-seat passengers are also becoming more popular.

◄ WHY ARE TRAFFIC SIGNALS NECESSARY?

Traffic signals control the flow of traffic. They keep it running as smoothly as possible and so help to prevent accidents.

If there were no traffic signals at junctions and in congested streets there would be chaos. Each motorist would try to barge ahead, and vehicles would get hopelessly entangled.

At a small crossroads, traffic signals may simply let traffic first on one road, then on the other move forwards. At busier spots, they can be far more complex. Green (for go) may show for longer on the main road than on side roads. Filters enable vehicles to turn right against the flow of traffic. At one moment in the sequence all the lights may show red (for stop), to allow pedestrians to cross safely. On main roads, a series of signals is often worked out so that if you travel at a given speed all the signals you pass will be green.

▼ WHY DO BICYCLE WHEELS HAVE SPOKES?

Spokes keep a bicycle wheel in shape. If it hits a stone, or goes down a pot-hole in the road, the spokes absorb the shock of the impact and prevent the wheel from buckling.

Most modern bicycles have between 24 and 40 wire spokes. They are hooked at right angles into a flange in the hub. At the other end they are threaded through a hole in the rim and kept in tension with a nut.

The first bicycles to have spoked wheels were the famous penny farthings, introduced in the 1870s. Although they look antiquated, even comical, nowadays, these were a great advance on previous bikes. They were all-metal and therefore very light, and the large front driving wheel made high speeds possible.

Many major cities have an underground railway network. The trains carry office workers, shoppers and so on between their homes and the city centre, by-passing the crowded streets above.

Underground railways are the only practical way of moving large numbers of people, most of whom are travelling at the same time of day. City centres

Bridges carry railways across roads, rivers or valleys. Tunnels take them through hillsides. In both cases, they avoid time-consuming detours.

Locomotives run most cheaply and efficiently on flat, straight track without curves or gradients. But this does not happen very often.

When they meet an obstacle such as a steep hill, railway

engineers have to decide whether to divert the line round it or to tunnel through it. Going round is probably cheaper, but will add to the journey time and so increase running costs. Tunnelling requires costly special equipment, but the result is a straight track. Trains do not have to slow down for curves and can therefore keep a higher speed.

Some natural features, such as the Alps, cannot be avoided. There is no choice but to build a tunnel.

would be completely choked if every commuter went by car or bus. Trains have only a few seats and a lot of standing room, so that a large number of people can squash in. At peak hours there may be a train every 1½ minutes.

Building a railway underground is extremely costly, so most lines run on the surface outside city centres. Modern planners try to develop mass transit systems. Trains run from the centre to the main suburbs and feeder bus services take people the rest of the way.

Regular maintenance of railway track is very important. Even the slightest crack in a rail could cause a serious accident.

Until recently, railwaymen would walk each stretch of track, tapping the rails to detect any faults and checking their alignment by eye. Nowadays, most of this work is done by machine.

Track-measurement cars, which run independently or as part of a regular train, look for faults. Other machines tamp (pack down) the ballast, check and correct alignment and replace sections of track.

One of the weakest and most dangerous points is the joint between two rails. Rails used to be short (no more than 30 metres), but since about 1960, welded rails have been laid on most lines. These are up to 400 metres long and when laid are welded (joined together) to form a rail several kilometres long.

▶ WHY ARE RAILWAY SIGNALS SO IMPORTANT?

Signals control the movement of trains. They keep them away from each other, and they also help to keep them running efficiently and on time.

The block is the basis of a signalling system. A block is a stretch of line in one direction. No more than one train may be in a block at one time. A stop signal controls entry into the block. If it is red, the driver may not proceed. A distant signal some way before the start of the block warns the driver of what the stop signal will show, so that if necessary he can start to slow down.

Keeping trains apart ensures two things: safety and efficiency. Electronic controls prevent a signalman putting two trains on a collision course. At busy stations and junctions, signals often give the speed at which the train is to go and direct the driver to a particular platform.

▲ WHY ARE SOME RAILWAYS ELECTRIFIED?

In most European countries, the busiest lines are electrified. Electric trains are cheap to operate, but the trackside equipment they need is expensive.

The advantages of electric trains are that the locomotives cost less to build and run than diesels. They take their power from outside, from the national grid, rather than creating it themselves, as diesels do. And electricity is cheaper than oil.

The big disadvantage is the cost of electrification. Overhead wires have to be installed along the entire line, and often bridges have to be raised. Electric locomotives are also less versatile: they can only run on electrified lines, whereas diesels can go anywhere. On the whole, heavily-used inter-city lines are worth electrifying, especially if the improved service makes trains competitive with planes. Some commuter lines are also electrified.

▼ WHY DO TRAINS RUN ON RAILS?

The wheels of every rail vehicle (such as locomotive, passenger carriage and freight wagon) sit on the rails. A flange, or lip, on the outside edge of each wheel holds it to the rail.

Steel wheels and steel rails are a very efficient combination. Because there is not much friction between them, a locomotive can pull ten times as much as a road vehicle of the same power.

Rails alone could not support the weight of a train. The rails are laid on steel or concrete sleepers. These sit on a bed of ballast, or stones, packed onto the roadbed.

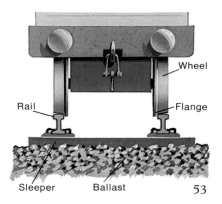

Wheel

Rail

Flange

Sleeper Ballast

▼ WHY WERE MOST EARLY AIRCRAFT BIPLANES?

Until the 1930s, most aircraft were biplanes. They had two sets of wings, one above the other. Struts and wires ran between the wings, holding them together.

When they built the first planes at the start of this century, the pioneers of flight had no previous experience to rely on. Cantilever wings, supported only by the plane's fuselage, were not to be developed for several decades.

Early aircraft wings were braced by a network of external struts and wires. These gave biplanes one of their main advantages, for it was possible to make comparatively thin wings which were efficient aerodynamically. In addition, wingspan (the distance from one wing-tip to the other) was shorter than on monoplanes, which have one set of wings.

The system of struts was also a disadvantage, for struts greatly increase drag. Drag is the resistance the air makes as a plane moves through it. Although drag was reduced by streamlining the struts and wires, it always remained greater than on an equivalent single-wing plane. Biplanes were also quite complex to manufacture and maintain.

On early aircraft, whether monoplanes or biplanes, the pilot sat in a tiny open cockpit well wrapped up against the weather. The first plane flew in 1903. Within ten years, air-mail flights had begun. Planes were used in World War I and by the early 1920s there were regular passenger services.

◄ WHY DO SOME AIRCRAFT HAVE PROPELLERS?

A plane's propeller uses the power produced in the engine to drive the aircraft forwards.

Propellers consist of three or four blades fixed to a hub. The blades act as aerofoils. As they turn, the air that passes over their top surface moves more quickly than air that passes underneath. As a result there is reduced pressure in front and increased pressure behind. This produces thrust.

Most propellers are variable-pitch: the angle of the blades on the hub can be altered to suit the plane's speed.

Propeller craft are still cheaper than jets at speeds up to about 800 kilometres per hour. Light planes are driven by piston engines, but larger propeller craft have turbo-props. In turbo-props air is compressed in a compressor, mixed with fuel and ignited in a combustion chamber. The exhaust gases drive a turbine that in turn drives both the compressor and the propellers.

▼ WHY ARE PASSENGER AIRSHIPS NO LONGER USED?

Airships were a popular means of long-distance travel in the 1920s and 1930s. But a series of accidents put them out of favour.

Airships work on the same principle as balloons. Most passenger airships were rigid: their envelope was surrounded by a rigid aluminium framework. Inside the envelope were gas bags filled with hydrogen or helium. To climb, ballast (normally water or sand) was released. To descend, gas was allowed to escape from the bags. Passengers and cargo travelled in a gondola beneath the envelope.

Four major airships crashed in the 1930s and many people died. Since then, airships have hardly been used at all. Now, however, there are plans to build cargo-carrying airships. These would be cheaper to build than ordinary planes and less costly to use.

▲ WHY DO HOT-AIR BALLOONS FLY?

Hot-air balloons use the difference in weight between hot air inside the balloon and colder air outside to obtain lift.

Ballooning is based on the theory of displacement. A container filled with hot air, or with a gas such as hydrogen or helium, will stay airborne. This is because the weight of the air it displaces is not less than the weight of the gas or hot air.

A hot-air balloon consists of a wicker basket in which the pilot and crew stand, a gas burner and a large airtight nylon envelope with a big opening at the base. To ascend, you turn on the burner, which heats the air inside the balloon in only a few minutes. To descend, you allow the air inside the balloon to cool. This takes quite a long time, and so the burner may be turned off for a while before the balloon starts to descend. On landing you open a ripping panel to deflate the balloon quickly and prevent the wind dragging it.

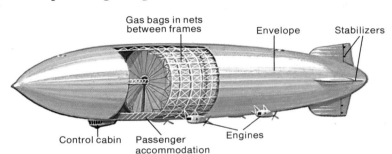

Gas bags in nets between frames — Envelope — Stabilizers — Control cabin — Passenger accommodation — Engines

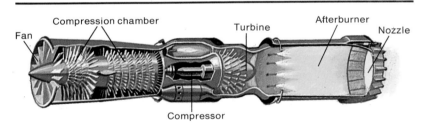

Compression chamber — Fan — Turbine — Afterburner — Nozzle — Compressor

▲ WHY DO SOME AIRCRAFT HAVE JET ENGINES?

Jet engines burn a mixture of compressed air and fuel. The exhaust gases stream out of the back of the engine and provide thrust to move the plane.

Most modern jet engines are turbo-jets. A fan at the front of the engine sucks air back to the compressor, which is driven by a turbine.

The compressor consists of a series of blades that compress the air (raise it to a very high pressure) and force it into the combustion chamber. Here it is mixed with fuel. The exhaust gases from the mixture drive the turbine that drives the compressor. The gases then stream out of the engine through a nozzle.

It is the thrust of the hot gases rushing out of the jet that propels the plane. An afterburner between the turbine and the nozzle acts as a second combustion chamber, increasing thrust. Vanes fitted to the exhaust nozzle provide reverse thrust.

▼ WHY DO YACHTS HAVE DIFFERENT TYPES OF SAIL?

Yachts travel by wind-power. Different sails are used in order to capture the wind and make full use of it.

The standard sails on a small yacht are the mainsail and the jib. The mainsail is a large tri-angular sail hoisted tight to the mast. Its bottom edge is fixed to a flexible boom. The boom follows each change of wind direction so that the wind always fills the mainsail completely. The jib billows out from the bows (front) of the yacht to catch the wind.

Many yachts also have a foresail (also hoisted from the mast), a cruising chute, used to catch the wind when it is behind the boat, and a spinnaker, which has a similar purpose.

Together, the jib and main-sail create an aerofoil shape that helps to propel the boat. Sails are continually adjusted to make best use of the wind.

▼ WHY DO BOATS HAVE KEELS?

The keel is the lowest part of a boat's bottom. It runs the whole length of the boat, and the rest of the vessel is built up from it.

The original purpose of a keel was structural. On wooden ships, the ribs (long wooden planks that formed the frame-work for the sides) were fixed into the keel. So too were the stempost (at the bows) and the sternpost. The keel itself was a long piece of timber, or several pieces joined together.

Modern sailing boats have fixed or drop keels. Fixed keels are ballasted to help overcome wind-pressure and keep the boat upright. Small yachts and dinghies have drop keels. In a following wind (blowing in the boat's direction of travel), the keel is raised to reduce resistance and increase speed. In a side wind it is lowered to resist drift and help the wind propel the boat.

▼ WHY ARE THERE DIFFERENT TYPES OF MERCHANT SHIP?

Ships are built to do different types of work. Cargo vessels vary according to the cargo they carry.

Ships are expensive to build and run. To make a profit, shipowners have to specialize. An oil tanker, for instance, is built to carry oil and nothing else. The engine room, bridge and crew quarters are situated at the stern, which leaves the rest of the ship for huge oil tanks.

Container ships are designed to carry as many containers as possible, both in the hold and on deck. Some general cargo vessels are still at work. Their holds can take virtually any type of freight.

Roll-on/roll-off ferries are an efficient way of carrying cars and their passengers. The cars are driven directly on board.

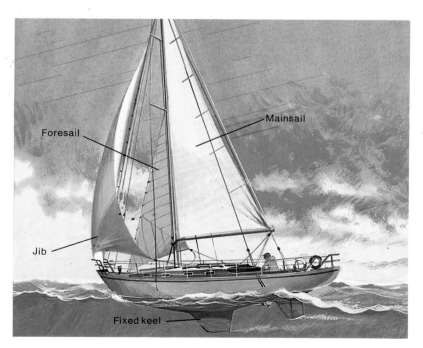

Foresail

Mainsail

Jib

Fixed keel

Oil tanker

General cargo vessel

Roll-on/roll-off ferry

Container ship

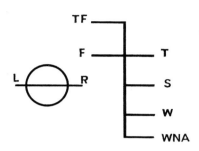

▲ WHY DO SHIPS HAVE LOAD-LINES?

Load-lines must be painted amidships on every merchant vessel. They are a safety measure that shows how much cargo a ship may carry.

The circle on the left is known as the Plimsoll Line. Samuel Plimsoll was a 19th-century social reformer who fought for greater safety at sea.

The horizontal line through the centre shows the depth to which a vessel may be loaded in summer in salt water. The initials *LR* stand for Lloyd's Register, the London organization in charge of enforcing these regulations.

The marks on the right show the maximum loading depth in different conditions, from bottom to top: Winter North Atlantic (WNA), Winter (W), Summer (S) and Tropical (T). The rougher the sea, the less cargo a ship may carry. TF and F stand for Tropical Fresh and Fresh. Fresh water is much less buoyant than salt, and so a vessel can safely be loaded to a greater depth at an inland port. Maritime charts show in which parts of the world each load-line applies.

▲ WHY WERE SUBMARINES INVENTED?

A vessel that can travel underwater is extremely useful in wartime. It can attack enemy ships and can also be used to land men and equipment in secret.

At the end of the 19th century submarines with their own weapon, the torpedo, were developed.

Submarines played a very important role in both world wars. They attacked enemy ships and prevented them from reaching home with food and equipment.

Many submarines are nuclear-powered. The first, the American *Nautilus*, was launched in 1954. They can stay submerged for a very long time and are armed with long-range missiles that can be launched underwater. They are now a vital part of a country's defence system.

Undersea craft used for civilian purposes are called submersibles.

▼ WHY DO CANALS HAVE LOCKS?

A lock is like a step. It raises or lowers barges on the canal from one level to another.

Canals are artificial waterways. When they run through hilly country, they have to change level at a lock.

A barge travelling upstream enters the lock through a pair of lock gates, which close behind it. Sluices at the other end of the lock are opened to let in water, and gradually the water-level in the lock rises. When it is the same level as the canal beyond the lock, the farther gates are opened and the barge moves forward.

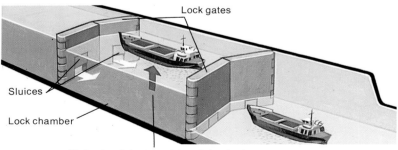

Lock gates

Sluices

Lock chamber

Water-level rises as sluices let water in

OUT IN SPACE

▼ WHY DO THE PLANETS REVOLVE ROUND THE SUN?

If the planets did not move, the Sun's pull would drag them inwards. But if they moved too quickly, they would fly off into space. The closer a planet is to the Sun, the faster it must move.

The planets all move in the same direction. They were probably formed from the same spinning cloud of material that produced the Sun. At birth, the Sun would have been spinning on its axis in a few hours. The cloud's pull slowed it down to its present 25-day period.

Apart from Venus and Uranus, all the planets spin in an anti-clockwise direction too. Venus spins very slowly backwards, while the axis of Uranus is tilted right over, so that it spins on its side.

▼ WHY IS THERE LIFE ON THE EARTH?

All life, from human beings to bacteria, will die if it becomes too hot or too cold, or if there is no air to breathe. No other planet in the Solar System has the right conditions for Earth-like life.

Known life is based on microscopic cells containing countless carbon atoms. But cells need substances containing elements such as hydrogen, oxygen and nitrogen if they are to survive. Although all the planets contain these basic elements, the Earth is the only planet with large amounts of water and oxygen. Both are essential for life.

The Sun's deadly rays are also filtered out by ozone (a kind of oxygen) in the upper atmosphere.

▼ WHY CAN'T WE SEE OTHER SOLAR SYSTEMS?

The nearest star is over a quarter of a million times as far away as the Sun. At this distance, even a large planet would be invisible in the biggest telescope. But some stars appear to 'wobble' slightly in the sky. This may be caused by the gravitational pull of a planet.

Most astronomers believe that a star in the constellation Cygnus known as *61 Cygni* has a planet revolving around it. Accurate measurements show that it is being pulled from side to side by an invisible companion, which must have about ten times the mass of Jupiter. It cannot be a faint star, because no star can be as small as this. It could not become hot enough to shine.

Mercury   Venus   Earth   Mars   Jupiter   Saturn   Uranus   Neptune   Pluto

A planet is a dark body orbiting a star. It can only be seen by the starlight it reflects. All the planets in the Solar System reflect light from the Sun. A moon (or satellite) also shines by reflection, but it revolves around a planet.

Of the nine planets, only Mercury and Venus have no known satellites. Spacecraft have helped to discover 14 around Jupiter and 23 around Saturn. Saturn has the largest satellite in the Solar System, Titan, which is 5120 kilometres across. It is larger than both Mercury and Pluto.

Most satellites are much smaller than their planet, although the Moon is a quarter of the Earth's diameter and Pluto is only twice as large as its moon Charon. Mars has two tiny moons, Phobos and Deimos, 23 kilometres and 13 kilometres across. Titan, and the four largest satellites of Jupiter, can be spotted with good binoculars.

▲ WHY IS THERE NO WEIGHT IN SPACE?

Weight is the force you feel when the floor, or a chair, stops you from falling towards the centre of the Earth. If a hole opened in the ground, you would be weightless while falling down it. This is known as 'free fall'. Flying through space, everything is in free fall.

An astronaut floats inside his cabin because he is moving at the same speed as the spacecraft. He will feel weight only if the motors fire, because the spacecraft will change speed or accelerate.

At take-off there is tremendous acceleration, and the astronaut is pushed down hard onto his couch. He feels several times heavier than he did when standing on the Earth's surface.

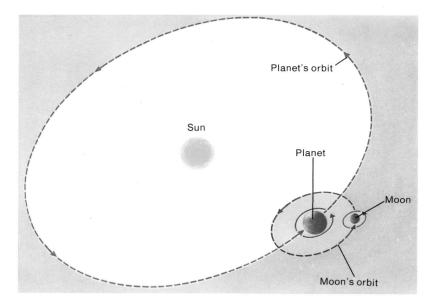

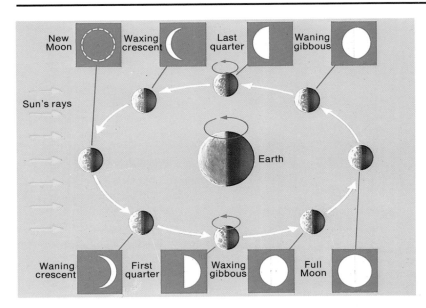

New Moon · Waxing crescent · Last quarter · Waning gibbous

Sun's rays

Earth

Waning crescent · First quarter · Waxing gibbous · Full Moon

◀ WHY DOES THE MOON WAX AND WANE?

The Sun can shine on only one half or hemisphere of the Moon. As it orbits the Earth every month, our view of this hemisphere changes. At New Moon it is turned away from the Earth, and cannot be seen, while at Full it is turned towards us. The time from New to Full is called 'waxing', after an old word meaning to grow. After Full, the Moon 'wanes' or shrinks to invisibility once more.

▲ WHY DOES THE MOON KEEP THE SAME FACE TOWARDS THE EARTH?

When it first formed, the Moon was probably spinning rapidly. But the Earth's gravitational pull has slowed it down. The Moon now turns on its axis in the same time that it takes to orbit around the Earth. This means that the same face is always pointing inwards, towards the Earth.

When the Moon was young and molten, the Earth's pull raised 'tides' in its interior, so that the dense rock below the lighter crust was forced into a slight egg-shape. The Earth pulled on this extra material, trapping the Moon into its present 'locked' state, so that we always see one side.

However, the Moon's surface does appear to swing or 'librate' to and fro by a small amount. This is caused by its elliptical orbit. It moves fastest when nearest the Earth, but its speed of rotation does not change, so the two get out of step.

Most people incorrectly call the thin crescent seen in the evening sky the New Moon. The only time when the true New Moon can be seen is when it passes across the Sun, causing an eclipse.

First Quarter, when the Moon has travelled a quarter of the way around the Earth, occurs a week after New, and the whole cycle of phases takes $29\frac{1}{2}$ days. In the crescent phase, the rest of the disc is often visible by sunlight reflected from the Earth. This is called 'Earthshine'.

▶ WHY DO ECLIPSES OF THE MOON OCCUR?

The Earth casts a long shadow into space, away from the Sun. If the Moon passes through this shadow, it grows dim. This can happen only at Full Moon, when it is opposite the Sun. However, since the Moon's orbit is tilted, it usually passes to one side of the shadow.

The Earth's shadow is not really black, since some sun-light filters redly into it, and

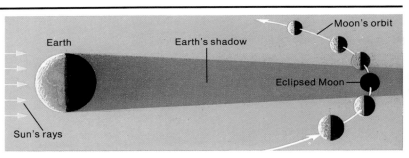

Earth · Earth's shadow · Moon's orbit · Eclipsed Moon · Sun's rays

the eclipsed Moon is often a deep copper colour. A total lunar eclipse can last for over one and a half hours, if the Moon passes through the shadow's centre. There is a faint outer shadow or 'penumbra' around the

central 'umbra', but the penumbra is hard to detect.

European observers will not have a good view of a total lunar eclipse until 10 December 1992. In the USA, a total eclipse will be seen on 17 August 1989.

▼ WHY IS THE MOON COVERED WITH CRATERS?

The planets and satellites of the Solar System were formed when much smaller bodies, a few kilometres across, collided together. Long after these worlds had taken shape, their surfaces were cratered by later collisions.

The lunar craters were probably formed during the first few hundred million years of the Moon's 4500-million year history. They are practically unchanged today, because the lunar globe cooled down quickly and preserved them. Also, there is no atmosphere to produce winds and weather, which could have worn these features away.

Similar collisions must have cratered the Earth at the same period in history. But the inside of the Earth has remained much hotter than that of the Moon. The Earth's crust is thinner, the surface has buckled and split, and the craters have been worn away by wind and rain.

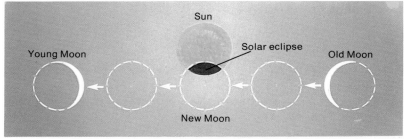

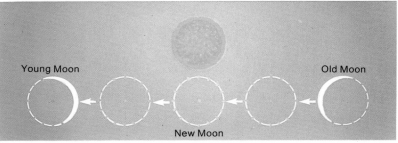

▲ WHY IS THE NEW MOON INVISIBLE?

The true New Moon lies almost between the Earth and the Sun. The Sun is so dazzling that nothing can be seen near it. (Never look directly at the Sun – it can damage your eyes.) Also, the night hemisphere of the Moon is turned towards the Earth.

The crescent Moon is usually first noticed in the evening sky about two days after New Moon. But some observers have seen very young crescents less than a day old, or very old crescents rising in the dawn less than a day before the next New Moon. The best time to look for a very young Moon is just after sunset in the spring (or, for an old Moon, at dawn in the autumn). Binoculars give you a much better chance of spotting it.

If the New Moon happens to be directly in line with the Sun, there will be a solar eclipse. But, since the Moon's orbit is tilted slightly, eclipses do not happen very often.

◄ WHY IS THERE NO LIFE ON THE MOON?

The life-forms that are familiar to us need air and water, as well as protection from some dangerous rays sent out by the Sun. The Moon has no atmosphere, no water on its surface, and it is completely exposed to space. Also, the temperature at midday is higher than that of boiling water.

Scientists have carried out experiments to find out if any living organisms could survive on the Moon's surface. They built special chambers in which these conditions were copied exactly. Even simple bacteria died.

However, in case some microscopic life-forms existed there, the first *Apollo* landing crews were immediately put into quarantine on their return to the Earth, and were kept under observation. This precaution was stopped when nothing harmful was found. The Moon seems to be totally lifeless.

If iron is left outside exposed to the weather, it turns rusty. The rocks on the surface of Mars contain iron which has turned into a kind of rust. This red, dusty material covers the planet's surface and is sometimes blown into huge dust storms.

Rust, or iron oxide, is a compound made of iron and oxygen combined together. There is now no oxygen left in Mars' atmosphere, since it also combined with other materials, forming water (with hydrogen) and probably carbon dioxide (with carbon). There are some 'rusty' rocks to be seen in various places on the Earth. One of these is the Grand Canyon in Arizona, USA.

On Mars, the dust is so fine that it hangs in the thin air as a permanent haze, causing the sky to be pink rather than dark blue. Sometimes the surface of Mars disappears from our sight for days or weeks beneath thick clouds of dust raised by the Martian winds.

▲ WHY IS MARS A DEAD WORLD?

The two *Viking* spacecraft which reached the planet in 1976 and examined its surface did not find any signs of life. They also discovered that the temperature is always below freezing point, and the atmosphere is thin and unbreathable.

All known living things contain cells made up of carbon atoms. The *Viking* landers did not find in the Martian soil any cells of the kind familiar to scientists, whereas the Earth's soil is teeming with them. Perhaps Mars has cells of a different type, or perhaps the spacecraft could have landed in more favourable places?

Mars may once have been warmer than it is now. From Earth, we can see channels in the surface which could have been carved by running water, and microscopic life-forms might have survived. But most people are convinced that Mars is a dead world.

▶ WHICH PLANET SPINS THE FASTEST?

Jupiter, which is the largest of the planets in the Solar System, spins on its axis in the shortest time, only 9 hours and 50 minutes. The next fastest planet is its giant neighbour Saturn, with a 'day' of 10 hours and 16 minutes.

Jupiter is much less solid than the Earth, and this causes different parts of its surface to rotate in different times. The Great Red Spot takes about five minutes longer to go round once than do objects near Jupiter's equator. The white oval cloud shown in the photograph, which has been seen and recorded for many years, also has its own different period of rotation.

When an object spins, an effect known as centrifugal force makes it begin to fly apart. This force causes Jupiter's equatorial regions to bulge outwards by about 5000 kilometres. If it did not rotate, Jupiter would be a perfect sphere.

▼ WHY IS VENUS SO HOT?

The atmosphere of Venus is about 90 times as thick as our own. Although it is always cloudy, enough sunlight breaks through to heat the ground during its four-year 'day'. This thick atmosphere acts like a blanket, holding in the heat, so that the midday temperature is 480°C, as hot as an oven turned fully on.

This is often called the 'greenhouse effect', since the air inside a greenhouse heats up for the same reason. Sunlight can pass through the glass, but when it has been absorbed by the ground inside, it is turned into heat radiation. This is blocked by the glass and cannot escape. Like the atmosphere of Venus (and of the Earth) it is a heat trap.

The atmosphere of Venus contains a lot of carbon dioxide, a gas which holds in heat very well. More carbon dioxide in our own atmosphere would raise the surface temperature of the Earth to a dangerous level.

▲ WHY IS IT DIFFICULT TO SEE MERCURY?

The innermost planet, Mercury, seems to swing out first on one side and then on the other side of the Sun. These appearances are known as 'elongations'. At eastern elongation (to the left of the Sun, as seen from the northern hemisphere) it is low in the western sky after sunset. At western elongation, it rises in the dawn sky. At these times, it looks like a star.

Mercury takes about 116 days to return to the same elongation, so there are three morning and three evening elongations each year.

Spring evening elongations and autumn morning elongations are the easiest to see, but Mercury is never very obvious to observers in Britain and northern Europe.

Mercury sometimes passes in front of the Sun. The next 'transit' of this type will occur on 13 November 1986.

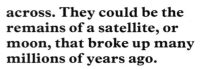

across. They could be the remains of a satellite, or moon, that broke up many millions of years ago.

◄ WHY DOES SATURN HAVE RINGS?

Nobody knows this for sure. The rings are made up of countless pieces of rock and ice a few centimetres or metres

Saturn's ring system is about 22 times the diameter of the Earth, but it is only a few kilometres thick. The particles are arranged in hundreds of rings, some of which seem to be twisted or braided together.

The brightest parts of the ring system probably contain more particles, and therefore reflect more sunlight back to us.

► WHY ARE THE GIANT PLANETS GASEOUS?

The Sun and the planets of the Solar System formed from a huge cloud of gas and tiny solid particles about 4600 million years ago. Most of the gas was hydrogen. Small planets such as the Earth lost a lot of hydrogen into space, but the giant planets kept all of theirs.

The pull of gravity of the giant planets is much stronger than that of the Earth. A person standing on Jupiter (supposing that it had a solid surface) would feel three times as heavy, and his legs would collapse beneath him. This pull of gravity prevents the hydrogen gas from escaping into space.

Hydrogen makes up about 90 per cent of the giant planets. Neptune, for example, probably has an outer shell of liquid hydrogen, an inner one of 'metallic' hydrogen (so called because electric currents can pass through it), and a small rocky core.

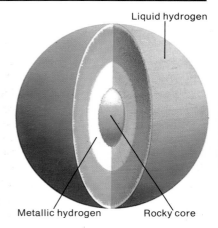

Liquid hydrogen

Metallic hydrogen　　Rocky core

◄ WHY WERE THE ASTEROIDS FORMED?

The asteroids consist of countless thousands of tiny planets left over from the Solar System's early history, when the Sun and planets were being formed. Most of them can be found in the wide gap between the orbits of Mars and Jupiter.

In the very early days of the Solar System, space must have been quite thick with bodies like these. Gradually,

though, they collided with each other and combined together to form the major planets. Later asteroid collisions formed the craters that can still be seen on Mercury and the Moon.

The largest asteroid, Ceres, is 1000 kilometres across, but most of the 2700 asteroids that have been observed are less than a tenth of this size.

Not all of these bodies orbit in the main asteroid zone. Some approach the Sun as closely as Mercury does, while the strange body Chiron is more remote than Saturn.

▶ WHY DO MANY COMETS APPEAR UNEXPECTEDLY?

Most comets take thousands or even millions of years to go round the Sun, and their orbits are so elongated or 'eccentric' that they travel far beyond Pluto. Very few become bright enough to see until they are as close as Mars. This is why some comets appear without warning every year.

Not all comets are surprise visitors. Some have small orbits and pass close to the Sun (perihelion) every few years. These can be predicted. The famous Halley's Comet, for example, which has a period of 76 years, has been observed regularly since 239 B.C. It will be seen again as it returns towards the Earth in 1986.

Each time a comet passes close to the Sun, it loses some material. So the really bright comets have long periods. Comet West, which was visible in daylight in 1976, will not return to the Sun for about a million years.

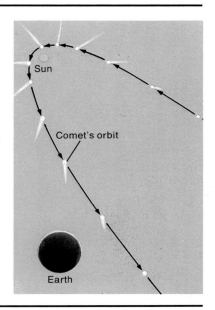

◀ WHY DOES A COMET HAVE A TAIL?

A comet is a crumbly body of rock and ice only a few kilometres across. When it passes near the Sun and becomes hot, the ice turns into gas, and dusty fragments are thrown out into space. Clouds of particles flying out from the Sun (the solar wind) push back this gas and dust to form the comet's tail.

Comets which have returned to the heat and wind of the Sun many times have lost most of their ice, and may produce only a dusty cloud or 'coma'. The finest tails are formed by comets that have passed the Sun only a few times. These can be millions of kilometres long, but are also very thin. In 1910, for example, the Earth passed harmlessly through the tail of Halley's Comet.

Some comets have only a single tail, but in 1744 a brilliant comet with six tails was observed! Most bright comets have two tails.

▶ WHY DO METEORS OCCUR?

A meteor is a streak of light high up in the atmosphere. It occurs when a tiny solid object, smaller than a marble, plunges through the air at a speed of many kilometres per second. The body is burned up by friction, leaving a white-hot trail.

Before entering the Earth's atmosphere, these bodies are called 'meteoroids'. They orbit the Sun like planets, some on their own, others in long swarms. These swarms are probably formed by comets. If the Earth's orbit passes through one of these swarms, a meteor shower is seen. On rare occasions, hundreds of meteors have been seen in a single hour.

A meteor shower occurs on the same date every year, when the Earth returns to the point on its orbit where it crosses the swarm. A famous meteor shower, the Perseids, is seen each year on the night of August 12–13.

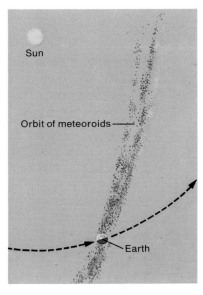

▶ WHY MUST YOU NEVER LOOK AT THE SUN?

The surface of the Sun is about four times as hot as a furnace. The lens or cornea in your eye acts like a burning-glass. If you look straight at the Sun, the lens will be destroyed for life.

Some people suggest looking at the Sun through smoked glass. Don't! The Sun may look dim, but the dangerous heat rays can pass through. Whenever there is an eclipse of the Sun, some people are blinded because they take foolish risks of this sort.

To observe the Sun, you should project its image

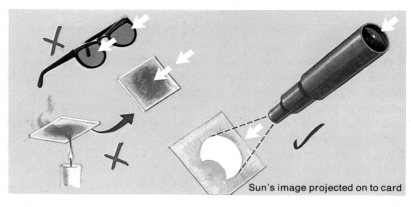

Sun's image projected on to card

through binoculars or a telescope onto a sheet of white card. Sunspots can then be seen easily and safely by a group of people together.

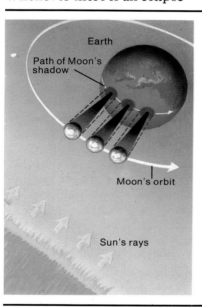

Earth

Path of Moon's shadow

Moon's orbit

Sun's rays

◀ WHY DO ECLIPSES OF THE SUN OCCUR?

The Sun is about 400 times the diameter of the Moon, but it is also 400 times as far away from the Earth. This means that both bodies look about the same size in the sky. If the Moon passes in front of the Sun, it can block out the brilliant disc, so that the faint surrounding 'corona' can be seen.

A solar eclipse can be seen only at New Moon, when the Moon is between the Earth and the Sun. But the line-up must be exact, or else only a partial eclipse will be seen, and the beautiful corona will not appear.

The diameter of the Moon's shadow on the Earth's surface, within which a *total* eclipse can be seen, is never more than a few hundred kilometres wide. This is why people wishing to see an eclipse must be prepared to travel a long way to see one. On average, a total eclipse is seen from the same site every 350 years.

▶ WHY IS THE SETTING SUN RED?

The Earth's atmosphere is like a pale red filter, and makes all the light coming from space turn slightly reddish. But an object which is very low in the sky has to shine through much more air than one high in the sky. This means that its light passes through more of this red filter, and its tint is deeper.

Since red light passes through

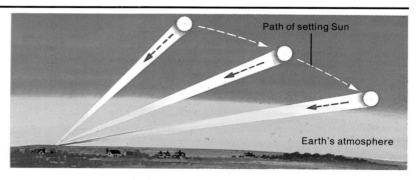

Path of setting Sun

Earth's atmosphere

air more easily than any other colour does, photographs of the Earth's surface taken from aircraft or satellites are often taken using a red filter.

The same is true of the thin atmosphere of Mars. Photo-graphs taken with a red filter show the planet's surface details well, whereas pictures taken with a blue filter show the hazy atmosphere of Mars and any thin clouds that may be present.

▶ WHY DOES THE SUN KEEP SHINING?

Even the tiniest object you can see with a microscope contains millions of atoms. Each atom contains much tinier particles still. Inside the Sun, atoms are being pulled to pieces and put together again in a different way. This gives out heat, and keeps the Sun shining.

The Sun is made up of 90 per cent hydrogen atoms, about nine per cent helium atoms (on the Earth, helium is a very light gas used to lift balloons), and one per cent other elements such as oxygen and nitrogen. In its centre, at a temperature of about 20 million degrees, hydrogen atoms are broken down and reassembled as helium atoms. Four hydrogen atoms (H) are required to make one helium atom (He). In this process, a burst of energy is given out.

The Sun's hydrogen will last for thousands of millions of years from now, at least as long as it has already existed.

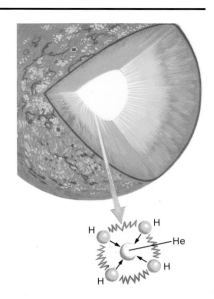

◀ WHY ARE THERE SPOTS ON THE SUN?

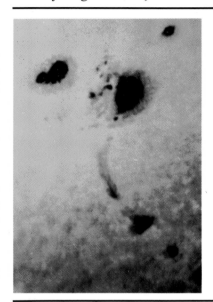

Sunspots are huge cool patches on the Sun's surface, caused by a very strong magnetic pull from the layers below. This magnetism acts rather like a cooling wind. Since they are cooler than the surrounding surface, sunspots appear dark.

Sunspots often occur in pairs or complicated groups, and may last for just a few days, or for many weeks. The middle of a spot is coolest and darkest, and is called the 'umbra'. Surrounding this is a lighter region called the 'penumbra'. The umbra is about 1000°C cooler than the Sun's surface, but it is still much hotter than some dim stars.

The Sun shows a regular change of sunspot activity. Every 11 years (as in 1969 and 1980) sunspots are especially common. A very big group of sunspots may measure ten times the Earth's diameter from one side to the other. One this size was seen in 1982.

▶ HOW BIG IS THE SUN?

The Sun measures 1,392,530 kilometres across, or 109 times the diameter of the Earth. Over a million Earths could be squashed into the Sun's globe. If the Sun were the size of a football, the Earth would be only two millimetres across.

Although the Sun is enormous compared with the planets, it is much smaller than many stars. The bright reddish star Betelgeuse in the constellation of Orion, is larger than the orbit of the planet Mars, or about 300 times the diameter of the Sun!

However, astronomers think that the Sun will eventually grow much larger than it is now. Its core will become so hot that the fierce radiation of energy will blast its outer layers into a glowing cloud, swallowing up the inner planets. It will become a 'red giant'.

◀ WHY ARE STARS INVISIBLE IN THE DAYTIME?

The stars are always in the sky, but the bright blue sky hides them. When the Sun sets, the blue fades away and the stars can be seen. This blue is caused by the atmosphere. In space, or on the airless Moon, the sky is always black and stars are always visible.

A powerful telescope can reveal some of the brighter stars even when the Sun is above the horizon. And the planet Venus is sometimes bright enough to be seen in broad daylight with the naked eye.

Some keen-eyed people have also seen Jupiter and Mars in the daytime! But the sky must be very transparent, without the slightest haze, and you must know exactly where to look.

When one of these planets is visible at dawn, it is interesting to see how long it can be kept in view, with the naked eye or binoculars, as the Sun rises and the sky turns to day.

▶ WHY DO STARS TWINKLE?

Out in space, stars do not twinkle. But when their light passes through the Earth's atmosphere, it is made to flicker by the hot and cold ripples of air. You can see this effect by looking at a distant view over a hot road in summer, or across a bonfire.

Astronomers call this effect 'bad seeing'. If the stars twinkle violently, their image

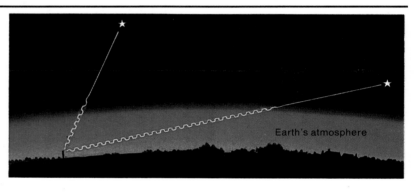

Earth's atmosphere

in the telescope will be a large blur instead of a tiny point, and small detail on the Moon or a planet will be invisible. Most bad seeing is caused by air currents several kilometres high. Large telescopes on mountaintops escape the worst of the unsteadiness, but bad seeing will also occur if the nearby ground is giving off heat waves.

◀ WHY ARE SOME STARS BRIGHTER THAN OTHERS?

There are two reasons. Some stars give out more light, like a large electric bulb compared with a torch. Also, some stars are much closer to the Earth than others are, and even a dim nearby star may appear brighter than a very luminous distant one.

This distance effect can be seen by looking down a lamp-lit road at night. The distant street lamps look much fainter than the closer ones do. Astronomers call a star's brightness its 'magnitude'. The *apparent* magnitude is its brightness as we see it in the sky. The brightest stars are about magnitude 0, while the faintest visible with the naked eye are about magnitude 6. A star's real brightness, or luminosity, is called its *absolute* magnitude.

Some stars are many thousands of times as luminous as the Sun, but they may be too far away to be detected without a large telescope.

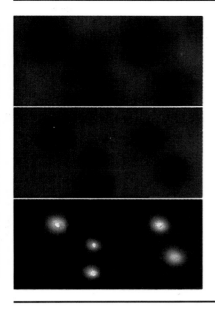

◀ HOW ARE STARS FORMED?

Stars begin their lives as very thin clouds of hydrogen gas, as shown in the top picture. As each cloud shrinks, the centre grows very hot due to the atoms of gas being squeezed together. Eventually, the clouds become so hot that they begin to shine as stars (bottom picture).

To begin with, the cloud which became the Sun was larger than the whole Solar System! Stars are usually formed in clusters, like the famous Pleiades, which were born about 100 million years ago inside a cloud or nebula several light-years across. The Sun was probably born in a cluster, but its companion stars have drifted away and cannot be identified.

Star-formation is still going on in our galaxy, since there are plenty of nebulae. But some other galaxies have very little gas left, and there are no new-born stars in their populations.

▶ HOW DO STARS DIE?

A star shines by turning its hydrogen into another element called helium. This change gives out heat, and keeps the star hot. When its hydrogen fuel runs down, the star begins to die.

Not all stars live for the same length of time. Small stars, dimmer than the Sun, use their fuel so slowly that they may have shone steadily ever since the Galaxy was formed, over 10,000 million years ago.

But very hot and brilliant stars, which contain a lot of hydrogen, may burn out in a few million years. Then they explode as a supernova, leaving wreckage like the Crab Nebula, shown here.

Others gradually blow their surfaces outwards like a red-hot cloud in the 'red-giant' stage. In the far future, the Sun may become a red giant.

▶ WHY DOES THE MILKY WAY LOOK PATCHY?

The Milky Way is our view of countless thousands of distant stars in the Galaxy. In some parts of the Milky Way, the stars are close together and the Galaxy shines more brightly. Elsewhere, it is hidden from view by huge dark clouds or nebulae that lie between us and the stars.

The Sun lies well away from the centre of the Galaxy, which is in the direction of the

constellation Sagittarius. The Milky Way is especially brilliant here, since the space in this direction is crowded with stars.

If we look in the opposite direction, towards Auriga, the Milky Way appears much fainter. The constellations Cygnus and Crux contain prominent dark nebulae in the Milky Way.

PLANET EARTH

◀ WHY DOES THE EARTH LOOK SO BLUE FROM SPACE?

A glance at a globe will give you the main answer: most of our planet is covered by sea. The Earth's atmosphere also gives a blue haze.

The oceans cover just over 70 per cent of the Earth's surface. All the Earth's light comes from the Sun in different kinds of light waves. Of the colours we see, red has a longer wavelength than blue.

Sea water and dust in the atmosphere absorb longer waves (red colours) but reflect shorter waves (blue colours). So the sea and sky look blue.

▶ WHY DOES A COMPASS NEEDLE POINT NORTH?

The Earth acts like a giant magnet, with its poles quite near the North and South Poles. The needle of a compass is magnetized, so it swings until one end points to the north magnetic pole.

The Earth's interior acts like a giant dynamo, generating its own magnetic field. This magnetic field is strongest near the poles. But the magnetic poles are not exactly the same as the Earth's poles, and they are not exactly opposite each other. They are also gradually moving. At present, a compass needle

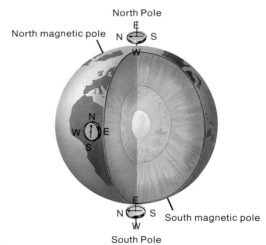

North Pole

North magnetic pole

South magnetic pole

South Pole

points towards the north of Baffin Island in northern Canada.

The Earth's magnetic field can be completely reversed, so that a compass needle would

point south. This has happened several times in the last few million years. The record of it is fossilized in some rock formations, like the lavas of the ocean floors.

▶ WHY DOES THE SUN RISE IN THE EAST?

Our Earth is rotating on its axis from west to east. As we travel eastwards on the rotating Earth, we pass into the zone of light where we can see the Sun. So we see the Sun first in the east in the morning.

Work it out for yourself with a globe and a torch. Shine the torch on the globe to represent sunlight, as in the diagram. Fix a model person, such as a little doll or a toy soldier, on to the globe so that it stands on the country where you live.

Now rotate your globe gently from west to east (see arrow on diagram). Watch the model pass from dark to light and light to dark. It is dawn as the model approaches the zone of light, and it will face east to see the Sun. It is evening as it passes into the zone of darkness and the Sun sets in the west.

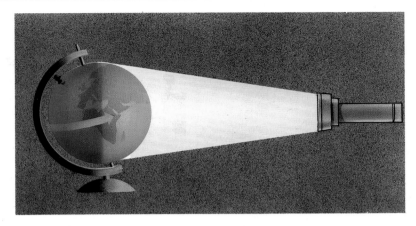

▲ WHY DO WE HAVE DAY AND NIGHT?

At any one time, half the Earth is facing towards the Sun and is in its light. The other half is shaded from the Sun, experiencing night. The Earth rotates on its axis once in 24 hours. During that time, the place where you live turns into the sunlight for part of the time (day) and then shadow (night).

The Earth's axis is an imaginary line joining the North and South Poles. This diagram shows the Earth in December. Notice how the axis is tilted. The South Pole is tilted towards the Sun (shown here by a torch) and it is summer in the southern hemisphere. Imagine the globe spinning: it would be light all the time in the Antarctic.

Towards the North Pole the days are short and in the Arctic it is night all the time. In June, the North Pole would be tilted towards the Sun and the Antarctic would be in the dark.

▶ WHY DO WE HAVE SEASONS?

In one year the Earth travels right round the Sun. The diagram shows the Earth at four different times of the year. Notice how the axis remains tilted at the same angle. First one hemisphere, then the other, is tilted towards the Sun.

In June, when the northern hemisphere is tilted towards the Sun, it is summer in Europe, Asia and North America. The Sun is overhead at the Tropic of Cancer. It is then winter in the southern hemisphere.

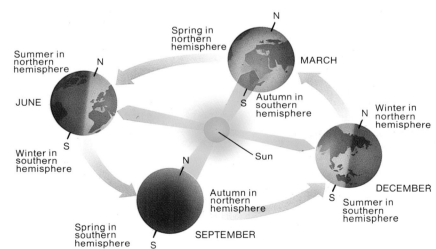

Six months later, in December, the southern hemisphere is tilted towards the Sun. So it is summer at Christmas-time in Australia, but it is winter in Europe.

The Sun is overhead at the Tropic of Capricorn. In March and September the Sun is overhead at the Equator. Both hemispheres are enjoying either autumn or spring.

**Fossils of the same kind
help geologists to put
different kinds of rocks in
the same age-group.
Fossils also give
information about the
conditions that existed
when they were alive. And
similar fossils in different
places have helped in the
study of continental drift.**

The fossils most useful to
geologists are of plants and
animals which originally lived
in a wide variety of places, but
only for a fairly short period
of time. These are called
index fossils. They prove that
the rocks in which they are
now found were all formed at
about the same time.

Fossils of plants and
animals that were very
sensitive to their environment
also tell us about the kind of
conditions existing when the
rocks were formed. Identical
fossils of plants and animals
which could not get across
water have been found on
opposite sides of oceans. This
means that such areas were
once joined before the
continents drifted apart.

▲ WHY DID THE DINOSAURS DIE OUT SO SUDDENLY?

**Dinosaurs became extinct
about 65 million years ago,
at the end of Cretaceous
times. They had existed
for 130 million years. No
one knows for sure why
they died out.**

Many other types of animals
died out at about the same
time as the dinosaurs,
including flying reptiles, sea
reptiles and the sea creatures
called ammonites. But a few
reptiles did survive from that
time and still exist today, such
as crocodiles, turtles and
tortoises, lizards and snakes.

There are many theories to
explain why the dinosaurs
died out, including changes of
climate and vegetation, cold
winters, and new parasites
and diseases. Volcanic dust,
meteorites, and even
mammals eating all the
dinosaur eggs, have also been
suggested. But the question
remains an unsolved mystery.

► WHY DO SOME ROCKS CONTAIN FOSSILS AND NOT OTHERS?

**Some rocks are too old to
have fossils. Other rocks
were too hot when formed.**

The three groups of rocks are
igneous, sedimentary and
metamorphic. Igneous rocks,
such as lava and granite, are
made from molten material
which cooled down. No plant
or animal could survive such
heat, so there are no fossils.

Sedimentary rocks are
made from material which
was deposited in water or on
land and many contain fossils.
Some, such as coal and some
limestones, are made almost
entirely of fossils. Others, like
sandstone, contain some
fossils.

Metamorphic rocks have
been changed by intense heat
or pressure. Marble is a
changed form of limestone,
for example. Any fossils in
them will have been changed
beyond recognition.

Sandstone (sedimentary rock) with fossil

Granite (igneous rock)

▶ WHY ARE THERE COAL SEAMS IN THE ANTARCTIC?

Coal is a rock made of the fossil remains of plants. The icy Antarctic has very few plants now, but when these coal seams were formed, there must have been plenty of plants. So the climate then must have been a lot warmer.

The coal seams of the Antarctic were formed in Permian times, about 250 million years ago. Modern theories of continental drift explain why Antarctica was much warmer then.

The maps below show where the continents could have been at different times in the past. In Permian times, Antarctica was probably part of the great continent called Pangaea. It was then at a much warmer latitude than now. When this huge continent split, Antarctica was part of Gondwanaland. When Gondwanaland split up, Antarctica slowly moved to the South Pole.

▶ WHY DO THE CONTINENTS MOVE ABOUT?

The crust of the Earth is a relatively thin layer. Scientists believe it is divided into huge sections called 'plates'. These are moved very slowly by convection currents inside the Earth.

The scientist Alfred Wegener suggested in 1915 that the continents might have moved. He had noticed that their shapes fit together like a jigsaw. His theory was called 'continental drift'.

These maps show the possible movement of the continents in the past. From a study of earthquakes and sea-floor spreading, scientists think that the continents and oceans are on the rigid plates into which the Earth's crust is divided. They can move because the rocks of the mantle (immediately below the crust) are moving.

Mantle rocks are not liquid, but 'gooey' like thick tar. Convection currents rise from the core, up through the mantle and spread out, moving the crustal plates like rafts.

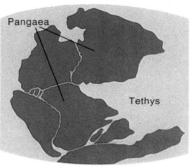

200 million years ago

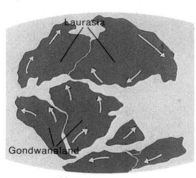

180 million years ago

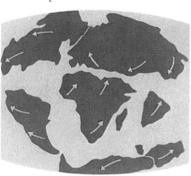

65 million years ago

◀ WHY HAVE SOME SEAS DISAPPEARED?

Small seas can be filled up by sediments brought in by rivers. Large shallow seas can also be partly filled with sediments, and their floors pushed up by earth movements.

An inland sea such as the Caspian is filling up quickly. Old maps show that it was much larger a few hundred years ago. Larger seas, like the Mediterranean and North Sea, can also fill up as sediments are carried in, and their sea floor may be pushed up by earth movements. Most of the sedimentary rocks found on dry land were originally laid down in seas that have long disappeared.

These maps show the pattern the continents may have formed in the past. Notice how the movements of the continents altered the shape of the oceans, such as Tethys.

Scientists believe that at the moment the Pacific is slowly getting smaller and the Atlantic wider. The Red Sea seems to be widening and may one day become a great ocean.

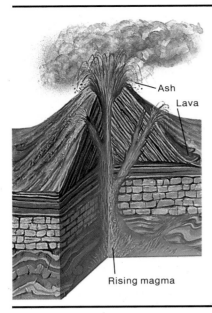

Ash

Lava

Rising magma

◄ WHY DO VOLCANOES ERUPT?

The lava of volcanoes is molten rock, most of which rises from the Earth's mantle. It is called magma. Some magma rises straight to the surface. Some is stored in a magma chamber in the crust, where the gases collect and help to drive the magma out.

The upper part of the mantle, under the Earth's crust, is nearly molten, or liquid. A slight rise in temperature or drop in pressure will make this magma melt. Because it is lighter than the rocks around it, the magma rises.

Magma contains several gases, including water vapour, carbon dioxide, sulphur dioxide and hydrogen sulphide. Bubbles of gas expand near the surface and drive the magma out as an eruption. Some gases burn as they reach the air and heat up the lava. Volcanic eruptions vary from place to place, mainly according to how fluid or gaseous the lava is.

► WHY DO EARTHQUAKES AND VOLCANOES OCCUR ONLY IN SOME PARTS OF THE WORLD?

The map shows where earthquakes have happened and where volcanoes have erupted. Both occur near the edges of the Earth's crustal plates.

The zones of volcanic and earthquake activity happen near the edges of the crustal plates. These are slowly moving together or apart. For instance, one zone encircles the Pacific. Here, the floor of the Pacific is being pushed down under the continents. This gradual movement of the Pacific plates against the continental plates causes earthquakes. The rocks of the sea floor are drawn deep down and heated up, and become the lava for volcanoes.

Another zone is the Mediterranean, where the African plate is pushing against Europe. One of the zones of greatest volcanic activity is the Atlantic, where new lava is added as Europe and America drift apart.

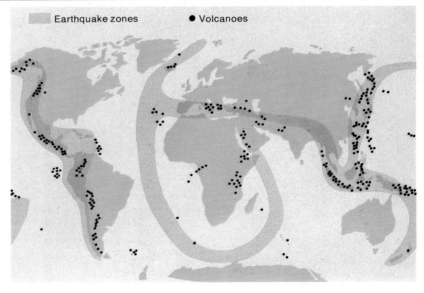

Earthquake zones ● Volcanoes

▲ WHY DO EARTHQUAKES OCCUR?

Earthquakes occur on deep faults in the Earth's crust. When the rocks suddenly move, shock waves are sent out in all directions. The movement is felt as an earthquake.

Many major faults in the crust are moving very slowly all the time. When the strain becomes too great the rocks suddenly move, causing an earthquake. The actual location of this movement is called the focus. This may be near the Earth's surface or deep in the crust. The place on the surface right above the focus is called the epicentre. Here the effects of the earthquake are felt first.

Scientists know why earthquakes happen and where they happen. The problem is to predict when they will happen. Delicate instruments can detect minor movements on a known fault which may build up into a major earthquake. But it is impossible to forecast an earthquake, or prevent it from happening.

▶ WHY DOES A GEYSER SPOUT HOT WATER AND STEAM?

All geysers occur in areas of volcanic activity, but not all volcanic areas have geysers. They only occur where water can soak through cracks in the rocks and collect under-ground.

Here the water is heated under pressure, and as it starts to bubble out there is a sudden gush of steam and hot water high into the air.

A geyser seems to work in a similar way to a pressure cooker. The higher the pressure, the hotter the water has to be to boil. Water deep underground in a volcanic area is super-heated before it reaches boiling point. When it boils, bubbles of steam rise, expand and push some water out at the surface. This reduces pressure below and lowers its boiling point. The super-heated steam and water is pushed out as a powerful jet. When enough water has seeped back and heated up, the process starts again.

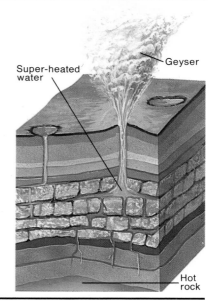

◀ WHY ARE THERE GREAT MOUNTAIN RANGES?

The highest mountains of the world are found in long ranges, like the Himalayas, the Andes and Rockies, the Alps and Pyrenees.

There are also great mountain ranges beneath the oceans, such as the Mid-Atlantic Ridge. All these mountain ranges are close to the edges of the great crustal plates, where one plate is pushing against another.

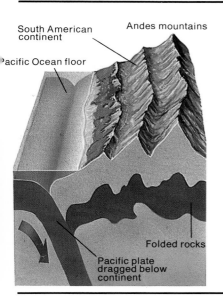

The diagram shows how the Andes are being formed. Part of the Pacific Ocean floor is being pushed under South America. Earthquakes and volcanoes result. And the rocks at the edge of South America are being folded and faulted and pushed up to form a great mountain range. Similar movements are happening along the western edge of North America.

Mountain ranges in other parts of the world have formed as two continents have moved closer together, such as Africa towards Europe.

▶ WHY DOES IT GET COLDER AS YOU CLIMB A MOUNTAIN?

As you climb up a high mountain the air gets thinner and colder. You may think you are getting nearer the Sun, but a kilometre is nothing when the Sun is 150 million kilo-metres away! In fact, the Sun heats the Earth, and it is the Earth that heats the air.

The Sun's energy that reaches the Earth's surface and heats

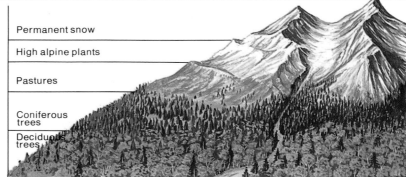

it up arrives as shortwave radiation. This is easily absorbed by the air. There is more air to absorb this heat near the Earth's surface than higher up. So the higher you

go, the colder it gets.

This has an effect on the vegetation that will grow at different altitudes in mountain areas, as the diagram shows.

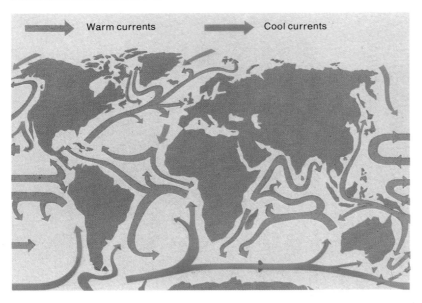

Warm currents →   Cool currents →

▲ WHY IS SEA WATER SALTY?

Sea water contains many minerals. They have all been washed out of the land and carried to the sea by rivers or glaciers. When sea water evaporates, these minerals remain in the sea and become more concentrated.

Salt (sodium chloride) makes up about 85 per cent of all the minerals in sea water. The saltiness of deep-sea water is fairly constant, but near the surface it varies from place to place. Saltiness is low where lots of fresh water is added from rivers and where there is little evaporation. The Baltic is one of the least salty seas. The Red Sea has few rivers and a lot of evaporation, and so is very salty. The saltiest seas are inland. The Dead Sea is the saltiest sea on Earth.

Salt can be made in hot countries by collecting sea water in shallow ponds and leaving the water to evaporate. The salt is left behind. But it is very difficult to make sea water fresh.

◄ WHY DO OCEAN CURRENTS OCCUR?

The main reason is winds. Prevailing winds blow the water at the surface of the oceans. But winds blow across land as well as sea. The ocean currents they cause do not stop or pile up against the land! So counter-currents develop.

Ocean currents are important because they move water from warm to cold areas, and from cold to warm areas. Currents that move away from the Equator are warm currents. Those that move towards the Equator are cool currents.

A cool current like the Labrador current, which flows from the Arctic to the Atlantic, may carry icebergs into the trans-Atlantic shipping lanes. A warm current like the Gulf Stream, which flows from the Caribbean to the north Atlantic, keeps ports in Norway free of ice all year.

Where these two currents mix, plankton flourish and fog is common in the famous fishing area of the Newfoundland Banks.

► WHY DO THE TIDES RISE AND FALL?

The water in the oceans is held close to the Earth by gravity. But the Moon and Sun also have some 'pull' on the Earth. The Moon's gravity affects water in the oceans. The sea is 'pulled' slightly towards the Moon, causing a bulge, or high tide. On the opposite side, the sea is pushed away, causing a second bulge.

High tides occur twice in about 25 hours. This is because at the same time as the Earth is rotating on its axis, the Moon is travelling round the Earth (every $27\frac{1}{2}$ days). So high tide is usually about one hour later every day.

The height of the tide varies. In the open ocean, the difference between high and low tide (the tidal range) is less than a metre. In an enclosed sea like the Mediterranean, it is much less, 30 centimetres. But where sea water is funnelled into a narrow bay or channel, the tidal range can be very high.

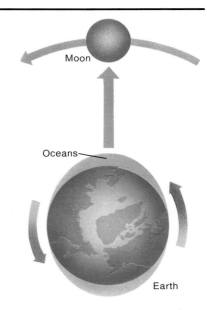

Moon

Oceans

Earth

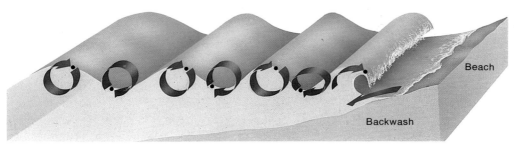

Arrows show movement of a water particle on the sea's surface

▲ WHY DO WAVES BREAK?

Winds cause waves, which form far out to sea. The wind pushes the water particles, which move round and round. Near the coast, where the sea is shallower, the sea bed interferes with the movement of the water. Then the top of the wave breaks on to the beach.

Waves have length as well as height. Wave length is the distance between one wave crest and the next. When the sea is shallower than half the wave length, the waves change. They drag on the sea bed and slow down. Then the crests crowd together.

On a gently sloping beach the crests often spill over and make a lot of surf. When the slope of the beach is steeper, the waves plunge on to the beach. They often have a strong backwash.

▲ WHY DO RIPPLE MARKS APPEAR ON SANDY BEACHES?

Both wind and waves affect the sand on the beach, causing ripples.

Ripples on sandhills are formed and move in a similar way to sand dunes. The ripples lie at right-angles to the wind and are very small. Wind blows the sand grains up the gentler face and they fall down the steeper downwind face.

Low tide ripples are formed by the gentle movement of small waves. Even with a very shallow sea and a gently sloping beach, the wind can form tiny waves. They work in the same way as larger waves but with much less power. However, they can mould the sand over which they pass. If the power of the waves increases, the ripples will be smoothed out.

▼ WHY DO SEAS SOMETIMES ERODE THE LAND?

The sea is very powerful, and attacks the land with waves, with stones and rocks in the waves, and with air compressed by the waves.

Erosion is at its most powerful when three conditions are combined: when the tide is high, the sea is rough, and the wind is on-shore. Waves can attack beaches, sand dunes and cliffs. The waves hurl stones picked up from the shore at the land. The waves also compress air in gaps and cracks in the rocks and force them apart.

A cliff face is only as strong as its weakest part. When the cracks are attacked, the other parts become weakened. Cliffs are undercut by the waves, as the diagram shows, and then cliff-falls occur.

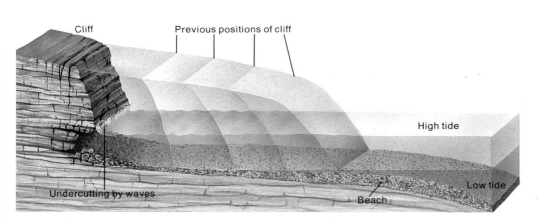

77

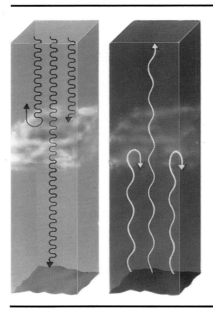

Our Earth is heated by the Sun. Clouds act like a blanket, so a cloudy day will stop some of the Sun's heat from reaching the Earth. But a cloudy night will stop the Earth's warmth escaping.

The left-hand diagram shows daytime, when the Sun warms the Earth. Clouds stop some of the Sun's heat (shown as red arrows) reaching the Earth. The right-hand diagram shows night-time, when there is no heat from the Sun. The Earth has warmed up during the day, and the clouds stop some of the Earth's heat (shown by white arrows) from escaping.

When there are no clouds, more of the Sun's heat reaches the Earth during the day. But during a clear night more of the Earth's heat is lost, so it can get very cold. In deserts it is hardly ever cloudy. Because of this, the days are hot but the nights can become very cold.

▶ WHY DO WINDS BLOW?

The Sun's heat warms the air and makes it move. This movement is called a wind.

Different parts of the Earth receive different amounts of heat, as shown in the top picture. Near the Equator, the Sun is overhead and heats the Earth intensely. Nearer the poles, the Sun's rays strike the Earth at a low angle so the heat is not so intense.

A lot of the Earth's heat is reflected back into space, by the atmosphere, clouds, dust in the air and by water, snow and ice. So some parts of the Earth are warm and some are cold.

Warm air tends to rise and creates areas of low pressure. Cold air tends to sink and creates areas of high pressure. As warm air rises, cold air flows in and replaces it. The greater the pressure difference, the stronger the wind blows.

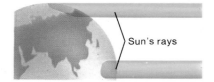

Sun's rays

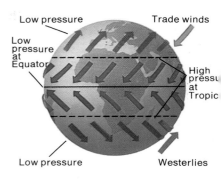

Low pressure

Trade winds

Low pressure at Equator

High pressure at Tropic

Low pressure

Westerlies

Winds blow in a curved pattern because the Earth spins, or rotates.

If the Earth stayed still, the winds would blow straight from an area of high pressure to an area of low pressure. But because the Earth is spinning so fast, winds swirl across the surface in a curved pattern.

In the northern hemisphere, winds blow into an area of low pressure in an anticlockwise direction. In this satellite picture, the winds are swirling anticlockwise, so this low-pressure system, or depression, must be in the northern hemisphere. A cyclone or hurricane shows the same swirling movement, but the pressure is very low and the winds are strong.

An area of high pressure is called an anticyclone. Winds blow out from an anticyclone. They move out in a clockwise direction in the northern hemisphere.

▼ WHY DO CLOUDS FORM?

A cloud is made up of tiny droplets of water or ice. When a cloud forms, the invisible water vapour in the air condenses into visible droplets of water.

All air contains water vapour. Warm air can hold more water vapour than cold air. If the air cools down, it cannot hold so much water vapour, and it turns into tiny droplets of water. You can see this happen when the hot steam from a kettle cools down in your kitchen. A lot of the water vapour turns into a cloud of water droplets.

Air cools down when it rises, because the higher in the atmosphere it goes, the cooler it gets. In the diagram below, the picture on the left shows air rising on a hot day above a city. The air rises and cools, and the water vapour condenses into droplets to form clouds. Towering clouds more than ten kilometres high can form in less than an hour on a very hot day.

Air rising over mountains (bottom right) will cool, and clouds will form. Clouds also form when warm, moist air rises over a layer of cold air.

Convection currents

Convectional rain

'Frontal' rain

Warm front

Air rises over hills

◄ WHAT MAKES IT RAIN?

When it rains, the tiny water droplets in a cloud form bigger drops which fall to Earth. No one knows exactly why this happens.

One theory about rain is that tiny water droplets stick together round something solid near the top of a cloud. This could be a speck of dust or an ice crystal. The larger, heavier drop of water then falls through the cloud, picking up more tiny droplets on the way. A raindrop may seem tiny, but it is several thousand times bigger than when it began as a tiny droplet.

► WHAT CAUSES FOG?

The cause of fog and cloud is the same, because fog is cloud at ground level. When damp air near the ground cools enough for the water vapour to condense into droplets of water, then fog and mist occur.

Air near the ground can cool for several reasons. If the surface of the Earth is especially cold, it will cool the air above it. If the air above a cold surface is almost still, then as it cools the water vapour will condense and fog will form. This can happen over a glacier, or over land which is cooling down quickly during a clear winter night (see top diagram). It can also happen over a cold sea. Sea fogs are common at San Francisco, in California, because there is a cold sea current near the coast.

Cold air is heavy and tends to sink. So fog may form on the cold tops of hills and mountains, and roll down into the valleys (see bottom diagram).

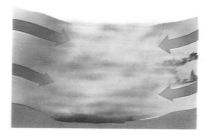

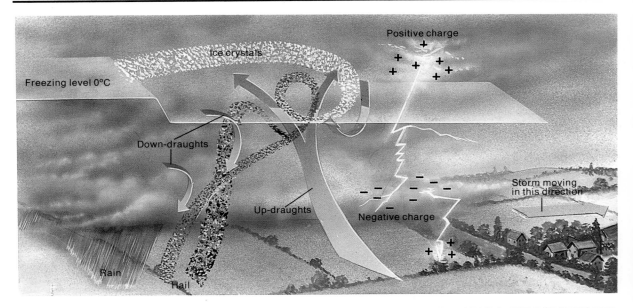

Positive charge

Ice crystals

Freezing level 0°C

Down-draughts

Up-draughts

Negative charge

Rain

Hail

Storm moving in this direction

▲ WHY DOES THUNDER ALWAYS FOLLOW LIGHTNING?

Lightning is a huge spark of electricity. Thunder is the sound made by the air as it is suddenly heated up by the lightning. They both happen at the same time. But light travels faster than sound, so we see the lightning first and then hear the thunder.

Thunder is caused when lightning heats up the molecules of air along its path. The heated molecules expand, collide with cooler molecules and set up sound waves.

Light travels very quickly, at about 300,000 kilometres per second. Sound is slower, travelling about 20 kilometres per minute. We see the flash almost as it happens, but we hear the sound afterwards.

The closer the storm, the closer together the flash and the thunder. Count the seconds between the lightning flash and the thunder. Divide this number by three to find out how many kilometres away the storm is.

▲ WHY DOES LIGHTNING OFTEN STRIKE TREES?

When lightning flashes from a cloud to Earth it takes the easiest path. It is always attracted to the highest point in an area, which is often a tree.

A storm cloud is like a giant electricity generator. The positive charge is at the top of the cloud and the negative charge at the base. Lightning is the spark between the two: first within a cloud, then from cloud to cloud, and then from cloud to Earth.

A lightning flash begins with a downward 'leader stroke'. This seeks out the easiest path to Earth. Air is very resistant to electricity and the leader stroke is attracted to the nearest high point. A single, damp tree offers a low-resistance route to Earth through the trunk.

Tall buildings also attract lightning, but fortunately they can be protected by lightning conductors. These are strips of copper (a good conductor of electricity), which offer a path for the lightning to reach the Earth.

▲ WHY DO WE SOMETIMES GET HAIL?

If you could cut a hailstone in half you would find that it is made up of lots of layers of ice. This is because a hailstone has risen high into the coldest part of the storm cloud again and again. Each time, another layer of ice freezes round it until the hailstone is so heavy that it falls to Earth.

The diagram shows a typical storm cloud. For a hailstorm, the top of the cloud must be above freezing level and there must be rapidly rising convection currents swirling up through the cloud. These carry up droplets which freeze near the top of the cloud and begin to fall. Before they reach the base of the cloud they are swept up again, and another layer of ice is added near the top of the cloud.

This process is repeated until the hailstone becomes too heavy for the up-draught to carry it higher, and it will fall to Earth. Hail from such storm clouds can form in temperate or tropical areas.

▶ WHY IS A SNOWFLAKE MADE UP OF CRYSTALS?

When water vapour cools, it usually condenses into water droplets. In very cold air the water vapour can condense directly into ice crystals. These crystals may cling together to make a snowflake.

Snow forms in a completely different way from hail. There is no freezing then thawing process, and no great up-draughts in a snow cloud. The temperature of the cloud is important. It must be cold enough for ice crystals to form straight from water vapour.

The shape of snow crystals varies according to the temperature and humidity of both the air in which they form and the air through which they fall. So it is not surprising that no two crystals are alike. 'Dry' snow falls in very cold, dry conditions. It has very small crystals and blows into snowdrifts easily. 'Wet' snow forms when the air is very moist and warm enough for the crystals to bond together.

▲ WHY DO HURRICANES CAUSE SO MUCH DAMAGE?

Hurricanes (also called typhoons or cyclones) are spiralling winds that can blow at 250 to 350 kilometres per hour. They start over a warm sea, absorb lots of moisture and hurtle towards the land. A hurricane can measure 400 kilometres across. It can uproot trees, destroy buildings and fling cars about.

The centre of a hurricane is called the 'eye'. It can be 40 kilometres across and is relatively calm. Around it swirl the hurricane winds. They are strongest nearest the eye. These winds circle round and rise up high in the atmosphere. Ocean water is drawn up into the eye, causing great storm surges and very high tides when it reaches the coast.

▶ WHY DO WE SEE RAINBOWS?

When sunlight passes through raindrops it is slightly bent. Sunlight is a mixture of colours. The raindrops bend some colours more than others, so they are separated out to make the colours of the rainbow.

Light rays travel in a straight line, but they do change direction when they pass through substances of different density, for example from air to water. Have you noticed that if you look down at a drinking straw in a glass of water it appears bent? This bending is called refraction. Raindrops refract sunlight.

Sunlight has to pass through raindrops at a low angle for the colours to show as a semicircular bow. This is why you see rainbows most often after showers in the early morning or late evening, and not at midday. From an aircraft or a mountain top you can sometimes see a rainbow below you as a complete circle.

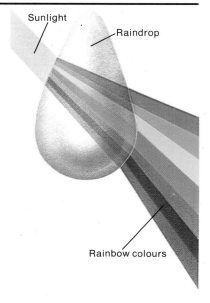

Sunlight

Raindrop

Rainbow colours

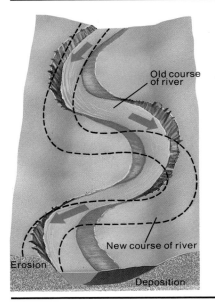

Old course of river

New course of river

Erosion

Deposition

◄ WHY DO RIVER
MEANDERS CHANGE THEIR
SHAPE?

The diagram shows that river currents hit the outside bend of a meander, where erosion occurs. On the inside of the bend, mud and sand are deposited. The two act together and change the shape of the meander.

Imagine you are in a canoe on the river. The current rushes you towards the outside of the meander. You paddle fast and just avoid hitting the bank. Then the current pushes you towards the other bank.

The current is constantly hitting the outside bank of a meander, undercutting it. When the river is in flood, it will wear away, or erode, the outside bank more quickly. The water will erode more powerfully when sand grains and pebbles are carried along.

One question remains: how did the meander begin? The slightest eddy will start a small bend in the river's flow. Once it has started, a meander gets bigger every year.

► WHY DO SOME RIVERS HAVE DELTAS?

A delta is shaped like a triangle, which is also the Greek letter D, called 'Delta'. A river delta is caused by deposits at the mouth of a river.

A river forms a delta if it transports more silt to its mouth than can be removed by waves, currents or tides. So deltas in lakes are very common. The Mediterranean Sea has a very small tidal range, so big rivers like the Nile (seen here from space) and the Rhône have very big deltas. Deltas are less common on rivers that flow to oceans and tidal seas.

When a river reaches the sea or a lake, the flow of water suddenly slows down. Still water cannot hold so much silt, and the river channel becomes blocked by deposits. The water spills over on both sides of the old channel. Soon the two new channels will be blocked, and the water will find new routes. So a delta has many water channels and is constantly changing.

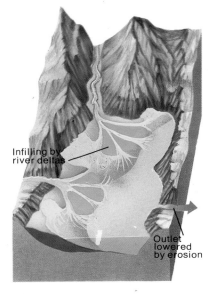

Infilling by river deltas

Outlet lowered by erosion

◄ WHY DO LAKES
SOMETIMES DISAPPEAR?

Lots of lakes have vanished. Some have been filled up, the water of others has flowed away, and other lakes in very hot countries have simply dried up.

Lakes can be filled up with material brought down by rivers. Deltas in a lake can eventually fill up the whole lake, which becomes an almost flat plain. Lakes in glaciated areas often vanish in this way.

Secondly, the outlet from the lake can be worn away so that the point at which the lake flows out is lowered. The lake will drain away, leaving dry land. Earth movements and volcanic activity can also make new outlets for lakes.

Some lakes dry up because of evaporation. Less water is being brought into the lake by rivers than is evaporating in the heat. After a series of wet years, Lake Eyre in Australia is a huge lake, but after several dry years it vanishes!

▶ WHY DO GLACIERS APPEAR?

A glacier is like a great river of ice. Glaciers appear when more snow falls than melts every year. The snow collects, squeezing the lower layers hard. It turns to ice and forms a glacier.

Glaciers appear only in cold climates where there is snow all year round. This is either in high latitudes (the Arctic or Antarctic) or at high altitudes (in mountains). Glaciers often appear first on the shady side of a mountain. There are more glaciers on the north-facing shady slopes of the Alps than on the sunny southern side.

Most glaciers today are left over from the last Ice Age. In the last two million years there have been five glaciations. Ice-sheets spread from mountain areas and from the Arctic to cover most of Europe. This happened when the climate was much colder than now. No one knows why the climate changed from warm to cold so many times.

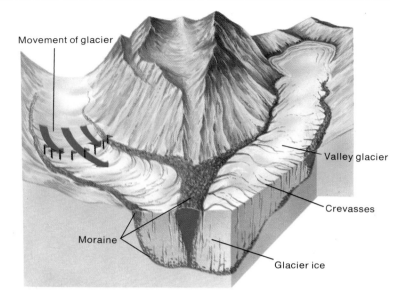

Movement of glacier

Valley glacier

Crevasses

Moraine

Glacier ice

▲ HOW DO GLACIERS MOVE?

All the time snow is being added at the top, the glacier will move forward under its own weight.

Glacier ice is under such pressure that the crystals melt slightly and can slip easily. The rate of movement will vary according to the slope of the valley, the thickness of the glacier, the roughness of the valley bottom and the temperature.

Movement can be observed by putting a line of stakes across a glacier. Those at the centre of the glacier move faster than those at the side. The differences in the rate of movement within a glacier create huge gaps or crevasses.

Rocks and stones (moraine) fall on to a glacier and are picked up by the ice at the bottom of a glacier. It is able to deepen and widen old river valleys by scraping the bottom like sandpaper and by plucking material out like pincers. The valley floors are deepened and flattened, and become U-shaped.

▶ WHY HAVE ICE-SHEETS ALTERED THE LANDSCAPE?

Ice-sheets alter a land-scape in very different ways from valley glaciers. They deposit moraine consisting of clay, sand and gravel.

Look at an atlas map of Finland or north-east Canada. There are thousands of lakes. These are examples of areas with very old hard rocks which have been eroded by ice-sheets. The material that is eroded is deposited on

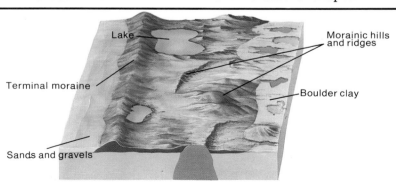

Lake

Morainic hills and ridges

Terminal moraine

Boulder clay

Sands and gravels

lower land where the ice-sheet is moving slowly or melting. In much of Great Britain and northern Europe, the lowland is a plain of boulder clay (clay with boulders) left by ice-sheets. They leave a landscape of low hills and hollows filled with lakes. At the edge of ice-sheets, great piles of rocks, gravels and sand are deposited. These make heath-covered hills such as the Lüneburg Heath in Germany.

▶ WHY ARE SOME SOILS FERTILE?

A fertile soil is rich in humus, bacteria and minerals. It also has enough water and a good texture.

The rock beneath soil is important. Permeable rocks, such as sandstone, make the soil light and dry. Impermeable rocks, like clay, cause heavy waterlogged soils. Very limey (alkaline) soils are bad for some plants. Broken-down rock fragments provide the minerals which plants need.

Climate and natural vegetation are also important. Plants need organic matter. Humus is made by soil bacteria which decompose, or rot, vegetation. They work best in a warm, airy soil, where there is plenty of leaf mould or grass. Water is very important in the right amount. Too much causes waterlogging or 'leaching', where the goodness is washed away through the soil. Too much evaporation can bring salts to the surface.

Top-soil (rich in humus)

Sub-soil

Weathered rock

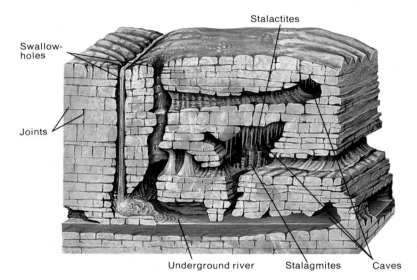

Stalactites

Swallow-holes

Joints

Underground river   Stalagmites   Caves

▲ WHY DO UNDERGROUND CAVES OCCUR?

The most spectacular caves are found in limestone. They are caused by rainwater and rivers dissolving the stone.

Caves in limestone are caused by water in two ways. Rainwater can dissolve the limestone and cracks are gradually enlarged to form caves. Also, underground rivers can erode the limestone. Many caves have rivers in them, or evidence of old, dry river beds. Some caves are made spectacular because of limestone deposits known as stalactites and stalagmites.

There are also caves in basalt rocks. These form when molten lava cools on the surface, but underneath it is still hot enough to flow away downhill. Caves remain when the lava has cooled down.

At the coast there are caves formed by the erosive power of the sea. With changes in sea level, they are sometimes found high above the present beach.

◀ WHY IS LIMESTONE DIFFERENT FROM OTHER ROCKS?

There are different kinds of limestone, all of which are sedimentary rocks. They are made of the fossil remains of sea creatures. You can see if rock is limestone by adding a drop of weak acid. Limestone will fizz and a little will dissolve.

In some limestone you can see the fossils of many shellfish. Other limestones are formed from corals which were living in warm seas millions of years ago. Some limestones were made chemically from calcium carbonate. Chalk is a soft type of limestone made from the remains of sea creatures.

All limestones are permeable. This means that water can pass through joints and cracks. All limestones will dissolve in rainwater, which is a weak acid. Some cracks become swallow-holes through which rivers vanish. Underground, water is continually dissolving the limestone and creating new caves.

◀ WHY DOES SOIL EROSION OCCUR?

Soil can be eroded, or worn away, by wind or by water. It occurs most often when a light soil is bare and dry and when there are strong winds or sudden heavy rainstorms.

Soil erosion is usually caused by bad farming. If a big field of light soil is ploughed, harrowed and left bare in dry weather, a strong wind will blow the topsoil away. Too many animals, such as goats, grazing on a semi-desert area, will eat plants and even roots and leave the soil bare. Then the wind will blow the topsoil away.

Water can cause as much damage as wind. Heavy rain or floods washing across bare earth can take the topsoil with it. Gully erosion is even more serious in some places. If a farmer ploughs down instead of across a slope, the furrows form little ditches down which rainwater can flow. In a sudden heavy storm, these ditches become rivers and dig deep channels.

▶ WHY DO SAND DUNES MOVE ACROSS DESERTS?

Only about one-fifth of hot deserts are covered with sand. Strong winds often blow in deserts. Because deserts are dry and do not have any vegetation, loose sand can be driven along by the wind. The constant movement of sand from one side of a dune to the other moves the dune.

The diagram shows the wind blowing across an area of crescent-shaped dunes called

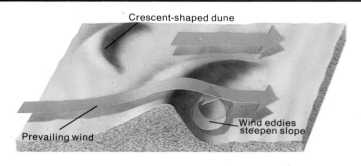

Crescent-shaped dune

Prevailing wind

Wind eddies steepen slope

'barchans'. The sand grains are blown up the gentle windward slope. They then fall over the crest and roll down the steep leeward slope. This slope stays steep because of wind eddies. The lower ends of the dune move faster, and are curved round by the wind in a crescent shape.

In other places, there may be long ridges of dunes, parallel to the wind direction, called 'seif' dunes.

◀ WHY ARE SOME DESERTS GETTING BIGGER?

Deserts are probably getting bigger because there is less rain. Also, strong winds move sand further out to cover the neighbouring land. Bad farming has also led to the soil being worn away, causing deserts.

Deserts have not always been the same size as now. There are prehistoric cave paintings of giraffes and elephants in the middle of the Sahara, and the Romans grew wheat in north Africa. The amount of rainfall in the Sahara and other deserts has probably changed. A small change can reduce the amount of vegetation and dry up rivers.

Very dry, hot winds can kill plants and move sand dunes. Dunes have overwhelmed villages and farms. Soil will blow away if land is over-grazed or ploughed up. Too many wells and pumps lower the level of water under-ground. This also affects the soil and makes irrigation more difficult.

THE PAST

◀ WHY DO HUMAN BEINGS WALK UPRIGHT?

Walking upright leaves our hands free to make tools and to explore our surroundings.

Human beings and the modern apes share common ancestors, who lived roughly 12 million years ago. These early man-like creatures lived in forests. They clambered about the trees, holding on with their arms. They had grasping hands, with fingers and thumbs. Their eyes were in the front of their heads, helping to judge distances accurately.

Two great changes then took place. The 'man-apes' moved out from the forests on to the grassy plains. They stood upright, walking on two legs. With their hands they used sticks as clubs and made simple pebble tools.

The first human being is known as *Homo erectus*, or 'upright man'. He was already one step ahead of all other animals. He could use his hands and his large brain. Our hands are marvellous tools, thanks to our walking upright.

▶ WHY DID PEOPLE START TO LIVE IN TOWNS?

When people first learned farming, they had to settle in one place. Gradually, villages grew into towns, protected by walls. This was the beginning of civilization.

The first civilizations were born in river valleys in Sumeria, India and China. Here people first learned to grow crops. They no longer had to roam from place to place in search of food.

In especially fertile spots, farmers could grow more food than was needed. There was surplus food to feed craftsmen such as potters, basket-weavers and builders. People began living in mud-brick houses, protected by walls. As more people moved into the towns, laws were made to

govern their affairs and look after them.

The world's oldest known town is Jericho, where there are remains of a walled town 9000 years old. In ancient Sumeria and India towns had shops, streets and drains. Trade and business flourished. Writing, mathematics and other skills were also developed.

▲ WHY WERE THE PYRAMIDS BUILT?

The pyramids were built as tombs for the pharaohs of ancient Egypt.

The early kings, or pharaohs, were buried in tombs inside roughly built stepped pyramids. The greatest pyramids came later. The largest is that of the pharaoh Khufu (also called Cheops) at Giza. Over two million blocks of stone were needed, each dragged into place by hundreds of slaves.

Inside the burial chamber, the pharaoh's body, preserved as a mummy wrapped in cloth, was buried in a stone tomb. Around him were food, clothes, weapons: everything that he might need in the next world.

▲ WHY DID THE GREEKS BUILD A WOODEN HORSE?

About 3000 years ago Greece and Troy fought a long war. In the end, the Greeks captured Troy by hiding soldiers in the wooden horse.

Troy (or Ilium) was a city in what we now call Turkey. After ten years of war, the Greek army laid siege to Troy but could not capture the city. Finally, the Greeks built a wooden horse, so big that soldiers could hide inside.

They left the horse outside the walls and the Trojans, unable to overcome their curiosity, dragged it into the city. During the night, the soldiers came out of the horse's belly and opened the gates. In rushed the Greek army, and Troy was lost. The story of the Trojan War, and of the wooden horse, was told by the Greek poet Homer.

▼ WHY DID HANNIBAL FIGHT A WAR WITH THE ROMANS?

Rome wanted to control all of the Mediterranean. Only Carthage, in North Africa, was strong enough to challenge Rome. And Carthage had a brave general called Hannibal.

Hannibal was born in 247 BC. By then Rome was already the strongest power in the Mediterranean. Carthage, Rome's chief rival, fought the Punic Wars against the Romans. Hannibal's own father drove the Romans out of Spain.

Hannibal led his army from Spain into France and across the mountains of the Alps. With him went his mighty war elephants. He attacked the Romans on their own ground in Italy and won great victories. But in the end, Rome's greater power proved too much and Hannibal had to return home. In 203 BC he was finally beaten at the battle of Zama, and was exiled from Carthage.

▼ WHY WERE THE EARLY CHRISTIANS PERSECUTED?

At the time of Christ, Rome ruled the world. Christianity seemed a threat to Rome, so the early Christians risked death for their faith.

Persecution of Christians began under the Roman Emperor Nero. He blamed the Christians for the fire that destroyed much of Rome in AD 64, and many were killed.

To escape arrest, Christians met secretly in underground catacombs, or burial

chambers. Persecution was at its height in AD 250 when the Emperor Decius declared himself to be a god. The Romans worshipped many gods. To believe in only one god was treason. Christians were burned alive or thrown to wild animals in the arena.

Despite the dangers, the new religion grew stronger. Finally, in AD 312, the Roman Emperor Constantine became a Christian. The Christians were able to worship freely and Rome became the centre of the Church.

▲ WHY DID THE CHINESE BUILD THE GREAT WALL?

Ancient China was a rich, proud empire. To keep out enemies, the Chinese emperors built a great wall across the north of China. It became one of the wonders of the world.

The Great Wall of China is 2400 kilometres long. Work on it began in 221 BC, linking a chain of forts along China's northern frontier. A road ran along the wall and signals could be sent quickly along it by patrolling soldiers.

Beyond the wall, to the north, lived fierce nomad tribes, such as the Mongols. For centuries, the wall kept out these terrifying barbarians. But in the end the Mongols did conquer China and made it part of their vast empire.

▲ WHY WERE THE VIKINGS BOLD SEA VOYAGERS?

In their fast wooden long-ships, Vikings crossed unknown seas to explore new lands.

The Vikings came from Scandinavia between AD 800 and 1100. They were farmers, but they were also fierce warriors and bold seamen. Their ships were strong and seaworthy. In them, Vikings sailed to Iceland and Greenland. One band, led by Leif Ericsson, crossed the Atlantic Ocean and founded a colony in North America. Another group, the Rus, settled in Kiev and gave their name to Russia.

Viking raids terrorized much of Europe. But in time, the Vikings gave up their roving and began to lead more peaceful lives.

▲ WHY WERE MEDIEVAL CASTLES BUILT?

In the Middle Ages, kings and lords built castles to defend their lands from enemies. Peasants from nearby villages took refuge inside the castles when war broke out.

The medieval castle developed from the wooden fort. At the centre of the castle was a tall tower, or keep. This was surrounded by high towers, thick stone walls and deep moats or ditches. The moats were filled with water. From the walls, soldiers could fire arrows at the enemy.

The windows were very narrow to stop missiles from entering the castle from outside. The attackers would lay siege to the castle, hoping to starve the defenders into surrender.

▼ WHY DID THE CRUSADERS FIGHT THE SARACENS?

For 200 years Muslims and Christians fought for the Holy Land of Palestine.

The Holy Land was sacred to both Muslims and Christians. In 1096, Pope Urban II called for a Crusade, or 'war of the Cross' to free the Holy Land from the Muslim Saracens.

The Crusaders came from all over Christendom, but failed to defeat the Saracens. The famous English king Richard the Lion Heart went to the Holy Land to fight and there was even a children's crusade in 1212.

The Crusaders brought home to Europe many new ideas from their wars in the East.

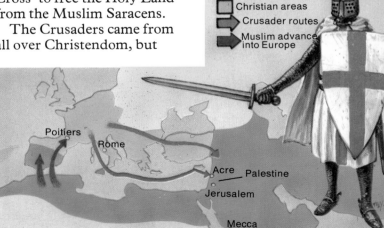

Muslim empire
Christian areas
Crusader routes
Muslim advance into Europe

Poitiers
Rome
Acre — Palestine
Jerusalem
Mecca

▼ WHY DID LUTHER DISAGREE WITH THE CHURCH?

The medieval Church was very powerful. Martin Luther, a German monk, dared to challenge it. His protest started the Reformation.

In Luther's day (he was born in 1483), the Church had fallen into bad ways. Some priests sold pardons for sins to anyone who could pay. The relics of dead saints, such as hair and bones, were also sold.

Luther thought this was

wrong. He pinned up a list of 95 arguments, setting out what he thought was wrong with the Church. For this rebellion, he was eventually excommunicated (cast out) of the Church.

However, others agreed with Luther. They set up breakaway 'Protestant' churches and refused to obey the Pope in Rome. In England, King Henry VIII made himself head of the Church. The movement begun by Luther is called the Reformation.

▶ WHY WERE THE FIRST AMERICANS CALLED INDIANS?

When Christopher Columbus set sail in 1492 he hoped to reach Asia and the rich spice islands of the Indies. Instead he found a New World and a new people, whom he called 'Indians'.

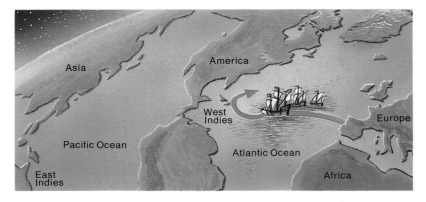

Columbus guessed that the Earth was round. By sailing west, he expected to reach the Indies faster than by sailing east, around Africa. He did not know that a huge continent, America, was in the way. So, when his ships reached the West Indies, he thought he must be in Asia and that the people living on the islands must be Indians. Many Indians were killed or enslaved by the Europeans who followed Columbus.

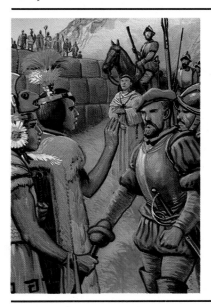

◀ WHY DID CHARLES I OF SPAIN BOAST: 'IN MY EMPIRE THE SUN NEVER SETS'?

King Charles I ruled Spain from 1516 to 1556. During his reign, Spain seized a huge empire in the New World. America's gold made Spain the greatest power in Europe.

As well as being King of Spain, Charles was also Holy Roman Emperor. He thought of himself as ruler of all Catholic Europe. He faced serious problems. Europe was being attacked by the Turks from the East, and was divided by religious quarrels.

But in the New World of America, Spain discovered vast riches. In 1519 Hernan Cortes and 500 *conquistadores* (conquerors) captured the Aztec Empire of Mexico. A few years later, Francisco Pizarro conquered the Incas of Peru. Treasure fleets bore gold and silver home across the sea to Spain.

In Europe's struggle for new lands overseas, Spain won a head start.

▶ WHY DID PEOPLE BEGIN TO WORK IN FACTORIES?

Until the 1700s most people lived and worked in the countryside. Then came a great change, the Industrial Revolution. People left their villages and went to work in factories in new, crowded towns.

Why did this change take place? One reason was the invention of the steam engine and other machines. Goods could be made faster by machine than by hand. Roads, canals and railways made travel much faster too. Jobs such as cloth weaving, which people had done at home, were taken over by machines.

Factories were built to house the machinery. And towns sprang up to house the workers who were needed to run the factories. Life for the workers was hard and dangerous. Britain was the first country to have an Industrial Revolution. Europe and America quickly followed, and by the mid-1800s factories were common.

▶ WHAT CAUSED THE FRENCH REVOLUTION?

In 1789 the French people rose against their rulers. They wanted freedom and equal rights for all.

France supported the American colonists in their fight for freedom from British rule. But France itself was divided. The king had great power, and the rich nobles and Church owned most of the land. The middle classes and the poor paid heavy taxes, but they had little power.

In 1789 the French parliament met for the first time in 164 years to discuss reforms. But the 'Three Estates' or groups (Nobles, Church, and Middle Classes and Peasants) could not agree. Fighting broke out in Paris when a mob attacked the Bastille prison. In 1793 the king was executed and France became a republic.

◀ WHY WAS ITALY UNIFIED?

In Roman times, Italy was one country. Later, it split up into small city-states. In the 1860s Italy threw off foreign rule and again became one nation.

The Italian city-states were once rich and powerful. But their quarrels made them weak, leaving them open to foreign conquest.

Napoleon Bonaparte tried to make himself king of a united Italy, but failed. The Austrians became masters of Italy. But the Italians dreamed of unity, and found a skilful statesman in Count Cavour and a bold general in Giuseppe Garibaldi.

In 1861 Garibaldi's one thousand 'redshirts' drove the Austrian forces out of southern Italy, and the struggle was over. Victor Emmanuel of Piedmont-Sardinia became the first king of a united Italy.

Only two states remained ouside the new kingdom. Venice joined Italy in 1866 and the Papal States, around Rome, joined in 1870.

▶ WHY DID EUROPEAN NATIONS COLONIZE AFRICA?

Until the 1800s Africa was the 'Dark Continent'. Few Europeans knew anything about it. As explorers began to uncover Africa's secrets, Europe took more interest in the huge continent's minerals and land. A scramble for colonies took place.

The journeys of Livingstone and Stanley in the 1860s and 1870s revealed Africa's vast wealth. The European nations of France, Britain, Belgium, Germany, Spain, Italy and Portugal rushed to seize as much land in Africa as they could. In 1885 they agreed to share Africa between them, and by 1902 Liberia was the only free African country.

Some European settlers, like the South African Boers, came to farm the new land. Others wanted minerals, such as gold and diamonds.

European officials were sent out to govern the colonies. The Africans lost their land and were often ill-treated.

▼ WHY DID RUSSIA HAVE A
REVOLUTION IN 1917?

**At the outbreak of World
War I, Russia was a poor,
backward country. The
war brought defeat and
discontent. In 1917 came
revolution.**

In 1917 Russia was a huge,
undeveloped country, ruled
by a weak Czar (emperor).
There was much unrest, and
the government could do
little. The army was defeated
in the war with Germany.
The people were starving.
There had already been one
attempt at revolution, in 1905.

Czar Nicholas II refused to

modernize his vast empire. In
1917 he was forced to
abdicate, but the new demo-
cratic government was weak.
Power was seized by the
Communists, led by Lenin,
who set up a revolutionary
government.

The Czar and his family
were executed. To mark the
change in Russia's history,
the capital was moved from
Petrograd (now Leningrad) to
Moscow. Russia became the
world's first Communist
state.

▲ WHY WAS WORLD WAR I
SO COSTLY?

**People called World War I
the Great War, the 'war to
end wars'. They were
shocked at the numbers of
people killed and injured.**

It was a mechanized war, the
first in which machine guns,
aircraft, submarines and
poison gas showed their
terrible power. In 1914 when
the fighting began, generals
on both sides expected quick
victories. Instead, the
German and Allied armies
became bogged down in
trench battles. Millions of
men died trying to gain a few
metres of muddy ground.

Cities were bombed from
the air. The war spread
overseas. Over eight million
troops were killed. More than
20 million more people died
from hunger and disease. By
1918 both sides were
exhausted.

▲ WHY DID THE CHINESE
MAKE THE LONG MARCH?

**In modern times, China
has seen civil war and
revolution. In 1934 the
founder of Communist
China, Mao Tse-tung, led
the Communists on an
epic 'Long March'.**

China's civil war was fought
between the Nationalist
government of Chiang Kai-
shek and the Communists.
Near to defeat, the Com-
munists had to retreat north
to Yenan in 1934. This
became known as the 'Long
March'.

In 1937 the two opposing
armies united to fight off the
Japanese. But in 1945 civil
war began again. This time
the Communists were vic-
torious. By 1949 they
controlled all mainland China.

The leader of the new
China was Mao Tse-tung, one
of the heroes of the Long
March. He ruled China until
his death in 1976.

After World War II Germany was in ruins. It was divided into East and West. The old capital, Berlin, was divided by a wall to stop East Germans fleeing to the West.

The victorious Allies divided Germany into zones. The Russians turned the eastern half into Communist East Germany. Many Germans fled to the west into the free Federal Republic of West Germany.

Berlin stands in East German territory. It, too, is

▲ WHY WAS THE UNITED NATIONS SET UP?

The most important problem today is how to settle international disputes peacefully. The United Nations was set up in 1945 with this as its chief aim.

During the 1930s the League of Nations could not stop ruthless dictators like Adolf Hitler. During World War II (1939 to 1945), the Allies fighting against Germany and Japan called themselves the 'United Nations'. When peace returned in 1945, they formed a new world organization called the United Nations. It had 51 members.

Today the UN has more than 150 members. It is not a 'world government', since its powers are limited. But it tries to persuade countries to solve problems peacefully, rather than by war.

▲ WHY WAS THE STATE OF ISRAEL FOUNDED?

The land of Palestine (Israel) is the ancient home of both Arabs and Jews. In 1948 a new Jewish 'homeland' was created.

Ever since the Romans occupied Palestine, Jews had dreamed of a new Israel. After Turkish rule of the area ended in 1918, both Arabs and Jews had claims to Palestine.

Jewish settlers began to emigrate to Palestine from all over Europe, where they had lived for centuries. Many more arrived after World War II.

In 1948 the United Nations decided that Palestine should be divided between Arabs and Jews. The state of Israel was created as a national homeland for the Jews. Since that time there has been conflict between the two peoples, and many Palestinian Arabs have fled to neighbouring countries.

now divided. Many people fled from East Berlin into the western part of the city, so in 1961 the East Germans built a wall across the city to seal the border.

The grim Berlin Wall is a constant reminder of the 'cold war' between East and West. Special crossing points are open only at certain times. People trying to escape to the West face armed guards and barbed wire. Those who are unsuccessful may be captured or shot.

HOW PEOPLE LIVE

▶ WHY ARE SOME BABIES BAPTIZED?

When a baby is baptized it becomes a member of the Christian Church. At the baptism service it is welcomed into the Church.

Parents who want their child to grow up as a Christian have it baptized. The ceremony is usually part of a normal church service, and often several children are baptized at the same time. It is a joyful event because the Church is welcoming new members.

The baby's family and friends gather round the font. The priest sprinkles holy water from the font over the baby's forehead, making the sign of the cross as he does so. He also names the child with the names its parents have chosen. (This is why a baptized person's first names are called Christian names.) Several people, usually relatives or family friends, promise to act as godparents. Their job is to make sure the child is brought up as a Christian.

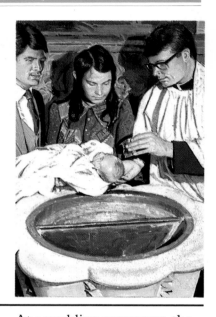

At a wedding ceremony, the couple promise to love and care for each other as long as they live. They formally accept each other as husband and wife and often exchange rings as symbols of their vows.

In most countries, people can choose between a civil or a religious ceremony. A religious ceremony often takes place in a church or chapel, but in some countries the ceremony may be held at the house of the bride, with a priest or holy man. It is also a tradition in some countries for the bride to bring a 'dowry', or her own wealth, for the groom's family.

Weddings are a time of great celebration and festivity. Guests bring presents for the bride and groom and wish them luck in their new life. The bride wears a beautiful dress that may have been specially made.

▲ WHY DO PEOPLE GET MARRIED?

In Western countries, a woman and a man usually get married because they love each other, and want to spend the rest of their lives together. In some countries, such as India, marriages are arranged by the bride and groom's parents. They choose a partner who they think is best suited to their son or daughter.

▶ WHY DO WE HAVE FUNERALS?

A funeral takes place soon after a person has died. The ceremony is a way of saying goodbye to the dead person. It enables people to show their grief at the death of a loved relative or friend.

Nowadays, many people are cremated. After a short ceremony, their body is burnt in its coffin. But burial in a cemetery is still common. The dead person's family and friends gather round the grave and the coffin is lowered slowly into it.

Funeral ceremonies differ around the world. For instance, in the Far East, the dead person's body may be placed on a funeral pyre and

burnt in the open.

In all societies, funerals are very important for the dead person's relatives and friends. They celebrate the person's life, and enable people to pay tribute to their memory.

▼ WHY DO CHRISTIANS TAKE COMMUNION?

The Communion service is the most important in the Christian Church. Worshippers share bread and wine that has been offered to the Lord.

The Communion service commemorates the Last Supper, when, on the night before He was betrayed, Jesus Christ shared bread and wine with His disciples. Each communicant at the Communion service takes a piece of unleavened bread and a sip of wine as well.

The Church offers itself and its own worship to God. He in turn accepts it and rewards the Church and its individual members with renewed life in Christ.

▶ WHY DO CHRISTIANS GO TO CHURCH ON SUNDAYS?

In Christian teaching, Sunday is known as the Lord's day. Christians go to church on Sundays to worship God.

The idea of a day of rest, when no work is done and you worship God, goes back to the Old Testament. According to the Bible story, God created the world in six days. He blessed the seventh day as a day of rest. The fourth of the Ten Commandments, which God gave to the Jews after they had been expelled from Egypt, told them to keep the Sabbath day holy and not to work on it. (The word 'sabbath' comes from the Hebrew word meaning to rest.)

The Sabbath is in fact Saturday, the last day of the week, and Saturday remains

the Jewish holy day. Muslims keep Friday as their holy day.

The Christians took Sunday, the first day of the week, as their day of worship. It was on Easter Sunday that Christ rose from the dead after He had been crucified.

Christians pray every day and try to live according to Jesus' teachings all the time. But Sunday, when most people do not have to work, is the day they keep for worship, and the most important church services are held then.

◀ WHY DO PEOPLE
CELEBRATE THE NEW
YEAR?

The New Year is a fresh start. You can put your mistakes behind you and begin again. People make New Year's resolutions, such as to work harder, or be kind to others.

New Year's Eve is a very popular time to have a party. As midnight approaches, the excitement increases and, as the new year begins, everyone drinks a toast.

In Scotland, Hogmanay, as the last day of the old year is called, is an important holiday. 'First-footing' is a well-known Scottish tradition. The first person to cross the threshold in the new year is warmly welcomed.

Throughout much of the world, the new year now starts on 1 January. But this was not always so. In England, it used to begin on 25 March until 1752. The Jewish New Year begins in autumn, on Yom Kippur (the Day of Atonement). The Chinese New Year, shown here, begins in February.

▶ **WHY DO CHRISTIANS OBSERVE LENT?**

Lent is the most serious period of the Christian Church's year. Many Christians put extra time aside for study and prayer. Some also give up a luxury, such as alcohol, and give the money saved to charity.

The 40 days of Lent run from Ash Wednesday to Easter. They recall the time of contemplation Jesus Christ spent preparing Himself for His ministry of teaching, healing and proclaiming the loving forgiveness of God towards mankind, as well as peace between people and their fellow creatures.

The Bible explains this time with a story. Jesus passed the 40 days in the desert, eating and drinking nothing. The devil tempted Him, offering Him money and power, but Jesus resisted. Christians try to make spiritual progress during Lent. They try, as Jesus did too, to resist inner temptations.

▶ WHY IS CHRISTMAS CELEBRATED?

Christians celebrate Christmas Day, 25 December, as the birth of Jesus Christ. But in most Western countries, Christmas is also an important national festival.

Christmas-time celebrates both a religious and a popular festival. According to the Bible, Jesus was born in a stable in Bethlehem near Christmas-time. The celebration of His birth is the second most important event in the Christian calendar, after Easter.

But there were mid-winter festivities long before Christ was born. People celebrated the winter solstice, when the days begin to lengthen and there is a promise of spring ahead.

Gradually, over the centuries, the two traditions came together. Nowadays, Christmas is a time of presents, decorations and feasting. It can be difficult to remember that it is a religious ceremony.

▲ WHY DO WE HAVE CHRISTMAS TREES?

In northern Europe, a tree is the central feature of Christmas decorations. People like to decorate their homes to celebrate the Christmas season.

Christmas decorations come from pagan mid-winter festivities. The Romans draped their temples with green branches The Druids used mistletoe, while the Saxons used holly, ivy and bay.

Christmas trees (which are usually young fir trees) came into use later. It is thought that Martin Luther introduced the tree lit with candles in the 16th century. But although Christmas trees were common in parts of northern Europe (where there are large forests of firs), they did not reach England until the 1840s. Queen Victoria married the German Prince Albert in 1840, and he introduced the idea. Christmas trees laden with presents and lights quickly became popular.

◀ WHY DO CHRISTIANS CELEBRATE EASTER?

Easter is the most important Christian festival. Christians remember Jesus Christ's death and His resurrection to a new life.

The Church's Easter celebrations are both solemn and joyful. The sober weeks of Lent prepare Christians for Good Friday. On this day, Jesus' crucifixion is remembered. He was nailed to the cross and left to die, as He Himself had prophesied.

After Jesus' death, His body was taken to a tomb. But on the third day of Easter, a Sunday, His followers came to the tomb and found it empty. Jesus had been resurrected, and had returned to his Father in heaven. This was a glorious moment for Jesus' followers, and it still is for all Christians. It is proof for them that He still lives and is real.

The Easter Day services in church are filled with joy and love and a sense of faith renewed.

▼ WHY DO MUSLIMS FACE MECCA TO PRAY?

Mohammed, who founded the Islamic religion, was born in Mecca. The city is now one of the holy places of Islam. Another is Medina, where he died.

Muslims pray to God five times a day: at dawn or just before sunrise, just after noon, before sunset, just after sunset and at the end of the day. To do so, they must face Mecca. In a mosque (the Islamic place of worship), the direction of prayer is marked by a niche in one of the walls, called the *mihrab*. On Fridays, an assembly replaces the noon prayers.

Another duty all Muslims must perform is to go on a pilgrimage, or *hadj*, to Mecca at least once. The pilgrimage takes place in the twelfth month of the Islamic year, which is about August. Pilgrims must make seven circuits of the *Ka'aba*, the shrine of the great mosque, shown here. This act is called the *Tawaf*. The ceremonies last for three days.

▲ WHY DO THE JEWISH PEOPLE CELEBRATE PASSOVER?

Passover is one of the main Jewish festivals. It celebrates the deliverance of the Jewish people from slavery. On the first evening of Passover, families gather together for a ritual meal.

Nearly 3500 years ago, the Israelites – as the Jewish people were then called – were the cruelly oppressed slaves of the Egyptians. Moses, an Israelite leader and prophet, led his people out of Egypt into the Sinai desert. He died just before they reached Palestine, which they believed was the 'promised land' God had given them.

The book of Exodus in the Old Testament tells the story of the Jews' escape in a very vivid way. God killed every first-born Egyptian, but He spared the Jews. He 'passed over' their homes. The ceremonies of the Passover meal recall the details of the story and remind Jews of their early history.

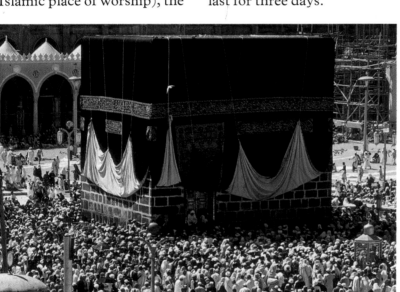

▲ WHY DO PEOPLE MAKE RELIGIOUS PILGRIMAGES?

People go on pilgrimages to the sacred places of their religion. The visit is a moving event in the pilgrim's spiritual life.

In the Middle Ages, many devout Christians went on pilgrimages. They dressed in pilgrims' clothes: a grey cloak, belt and hat, wore a cross around the neck, and carried a staff, sack and gourd. Thousands of pilgrims made their way across Europe to the great shrines of Rome and Assisi, Santiago, Chartres, Cologne, Aix-la-Chapelle and Canterbury.

They wanted to honour their God and to come closer to Him through prayer, to seek assistance, especially the cure of illness, and to seek forgiveness for their sins.

Modern pilgrims travel for the same reasons, but they no longer go by foot along the old pilgrim routes. Believers of many other faiths also make pilgrimages. Muslims visit Mecca and Hindus visit the holy city of Varanasi.

Ganesh

Kali

Shiva

▲ WHY DO HINDUS WORSHIP MANY GODS AND GODDESSES?

Hindus believe that God is present in everything. Whichever god they worship, their prayers will reach the same divinity eventually. The gods represent different qualities and features of Nature.

Hinduism is a major religion of India. The three main gods are Vishnu the preserver, Shiva the destroyer, and the wife of Shiva, a goddess known by many names such as Parvati or Kali. Each god is a part of Brahma, the universal spirit.

Worshipping one of these gods does not mean rejecting the others. Each god has several different forms. Shiva, for instance, is the lord of spirits. He also protects cattle and is the god of letters and music. In some parts of India he is also worshipped as the god of dancing.

Less important gods are found in hills, woods and streams, trees and plants, animals and snakes.

▼ WHY DO HINDUS AND BUDDHISTS MEDITATE?

Meditation plays an important part in Eastern religions and philosophy. Meditation affects both the mind and the body. It aims to control the person's body, improve character and help to develop wisdom.

In Hindu philosophy, meditation takes the form of yoga. The word yoga means 'diversion of the senses from the external world, and con-

centration of thought within'.

In Buddhism, meditation forms an important part of the 'Noble Eight-fold Path' that disciples must follow to understand the truth of their religion. Meditation starts with simple breathing exercises that aim to produce control of the body. When this is achieved, the character can be perfected, and finally wisdom and intuition can develop. Nowadays, many Western people find that meditation helps them to cope with the pressures of daily life.

▼ WHY DO MOSQUES HAVE MINARETS?

A minaret is the tower of a mosque. From the top of the minaret, five times a day, the *muezzin* calls Muslims to prayer.

The call to worship, or *azan*, follows a set pattern. The *muezzin*, or crier, who is an official of the mosque, stands in the minaret and repeats the same phrases in the same order, the same number of times. He must face Mecca and keep the points of his

forefingers in his ears. The *azan* must be heard with reverence. Work should stop, as should people in the street. Devout Muslims reply to each of the *muezzin*'s phrases.

There were no minarets when Mohammed, the founder of Islam, was alive. Nor were believers called to prayer in such an elaborate way. But the Jews called their people together with a trumpet, while the Christians used a bell. So Mohammed ordered one of his followers to climb to the highest roof to summon people to prayer.

◀ WHY DO SOME MUSLIM WOMEN WEAR VEILS?

In some Muslim countries, women cover their faces with a veil whenever they go outdoors. The laws of their religion say that men may not see their faces. Some Muslims believe that women should not take jobs or play an active part outside the home.

In traditional Muslim society, women lived separately from men. At home, or outdoors, they wore a veil and a long black cloak if men were present. They could appear without a veil in front of their husband, father or brother.

While men lived on the ground floor of their houses, women's quarters were on the upper floor and they had a separate entrance, courtyard and garden. But even though they lived apart, women had strong legal rights and could own money and property.

In many Muslim countries, these traditions are now dying out. But they are still found in some, such as Saudi Arabia.

▶ WHY DO THE SCOTS WEAR TARTANS?

Every Scottish clan, or tribe, has its own tartan. A tartan design consists of stripes of different widths and colours. These cross each other on a background of solid colour, to make a chequered pattern.

Kilts and plaids are the main items on which tartans appear. A kilt is a knee-length pleated skirt. A plaid is a long piece of cloth wrapped around

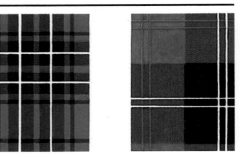

the chest and over the left shoulder and fastened with a brooch. Highlanders also wear a sporran, or pouch, which hangs in front of the kilt, a doublet, bonnet and brogues, or stout shoes.

After the 1745 uprising, when Scotland tried to regain independence from England, tartans were banned for 40 years. But when Queen Victoria became interested in them, they immediately became popular, and have remained so ever since.

▶ WHY DO SIKHS WEAR TURBANS?

A turban is a long silk or cotton scarf wound around the head. Male members of the Sikh religion may never appear in public without a turban.

The turban is worn to protect the head against the Sun's heat. But it also plays an important part in religious and social customs. Its size, shape and colour vary according to the rank and occupation of the wearer. A

turban decorated with a jewel, for instance, shows that the wearer has a very high rank.

Male Sikhs must swear to follow the five 'k's. These are *kes*, or long hair, which may never be cut and around which the turban is wound (Sikhs also never shave); *kach*, knee-length underpants; *kara*, iron bangle; *kirpan*, sword or dagger; and *kangha*, a comb.

The five 'k's had a military purpose, but they also have a religious meaning. *Kara*, for instance, stands for obedience.

▶ WHY DO PEOPLE USE COSMETICS?

Men and women have been using cosmetics for thousands of years to improve their appearance and to keep themselves clean and healthy.

Cosmetics were especially popular in ancient Egypt, as the picture shows. Beautiful jars and bottles held sweet-smelling oils and perfumes. People painted their faces with bright colours and coloured their lids, lashes and eyebrows black with *kohl*. This is still used as a cosmetic.

In some primitive tribes, people still paint their faces and bodies with bright colours and patterns, sometimes to show that they are of high rank. They use coloured pigments from the earth and from plants.

People still use cosmetics to make themselves look more attractive and to look after their skin and hair. Nowadays, cosmetics are made very cheaply from synthetic materials, vegetable oils and whale oil.

▲ WHY DO PEOPLE WEAR UNIFORMS?

Uniforms help us to recognize their wearers instantly. Everyone knows what their job is. Uniforms also make us respect the wearers' authority. They are also used for protection.

Some uniforms identify people doing a particular job. Airlines, for example, make their staff wear the same kind of uniform. Others also help to give authority to the person doing the job. Almost everyone behaves better when they see a policeman in the distance!

Military uniforms are used both for protection and to identify everyone. In a battle it is important to see who is fighting on your side.

People doing dangerous work need to be protected as much as possible. Firemen, for example, wear uniforms designed to protect them from heat.

◀ WHY DO PEOPLE LIKE JEWELLERY?

Jewellery makes people look attractive. Wearing an expensive jewel also shows that you are rich enough to have bought it.

Jewels have been worn for thousands of years. Nowadays, they are mainly used as decoration, but this was not always so. Many years ago, some kinds of jewellery, such as clasps and brooches, had a special use. They gradually became more and more richly worked and decorated.

Other kinds, such as necklaces, were thought to have magical properties and were worn to protect their owner. Some people still wear lucky charms today.

Expensive jewellery used to be worn only by the very rich. It consists of precious metals such as gold and silver, precious stones such as diamonds, or both. Today, machines make huge quantities of jewellery very cheaply, but the best pieces are still made by hand.

▼ WHY DO WE USE SOAP?

Soaps and detergents keep things clean. We use them to wash our bodies and also our clothes and dishes. They work by lifting dirt from greasy surfaces.

Soap is made by boiling animal fats or plant oils with soda to make crude soap. Brine is added to this and the resulting soap curd is heated. This produces hot, pure soap ready to be cooled and cut into bars or turned into products such as shaving soap.

Unfortunately, soap does not work well in 'hard' water, but detergents do. These are artificial soaps made from chemicals obtained from oil or petroleum. They are used a great deal both at home and in industry.

Many substances are added to crude soap to make it suitable for use as toilet soap. Coconut oil is added to make it lather quickly. Dyes, perfumes, water softeners and germicides, which are tiny substances that kill germs, are also added.

▲ WHY DO WE WEAR HATS?

Hats have a very important use in protecting people from the weather, and from injury in an accident. But they are also decorative and fashionable.

In an accident, a sudden blow to the head can be very dangerous. Workers in factories and on buildings sites should always wear industrial helmets to protect themselves. These consist of a hard shell on the outside, and a network of shock-absorbing straps on the inside. Motor cyclists' crash-helmets work in the same way.

There are many styles of hat that protect the wearer from rain, snow or cold. Hats worn for fashion may have no protective use. They can be large or small, elaborate or simple, and may be made from all kinds of materials – even fruit and flowers!

Perhaps the most versatile hat of all is the cowboy's 'ten gallon' hat. This shields him from the burning sun and serves as a pillow and water bucket as well!

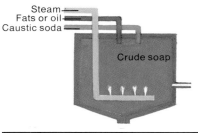

Steam
Fats or oil
Caustic soda
Crude soap

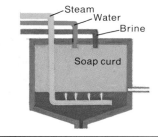

Steam
Water
Brine
Soap curd

Sperm whale

Musk deer

Civet

Beaver

Flower oils

Essential oils

Alcohol

▲ WHY DO WE USE PERFUMES?

People use perfume to make themselves smell sweet and more attractive. But in the days when standards of hygiene were poorer, perfumes helped to disguise many unpleasant smells.

The word perfume comes from Latin. *Per fumum* means 'from smoke', and the earliest makers of perfume burnt aromatic woods and gums. Their perfumes were used at religious ceremonies.

A perfume is the fragrance of flowers, leaves, stems, roots or even wood, turned into liquid form. This fragrance consists of tiny drops of essential oils, and it is the perfumer's art to extract these.

Modern perfumes are a mixture of real and artificial flower oils, alcohol and fixatives, which 'fix' the fragrance and make it last. Some ingredients are taken from animals, such as the musk deer, the beaver, civet and sperm whale.

▲ WHY DO WE WEAR SHOES?

Although shoes are made in many different styles and of many different materials, they are all designed to protect our feet from the weather and from rough ground.

The earliest shoes were sandals, made first of woven grass or papyrus and later of leather. These were ideal in hot countries, but useless in colder climates, and so shoes consisting of a soft upper part and a hard base developed.

Leather has always been the commonest material for making shoes, but other animal skins are also used, including pig, crocodile and even snake. Synthetic materials are also used.

Shoes, and boots that extend to the knee or beyond, are fashionable as well as practical. They have been made in an enormous variety of styles: decorated and plain, wide-toed and pointed, high-heeled and low-heeled. Today, most shoes, except very expensive ones, are made by machine.

▼ WHY ARE SOME PEOPLE TATTOOED?

A tattoo is a coloured design on a person's skin. It is made by placing colouring material in small deep holes pricked in the skin. Tattooing is usually done to decorate the body.

In the Far East and the Pacific, tattooing is an ancient art. (The word 'tattoo' comes from Polynesia.) But it was unknown in the West until the 18th century, when sailors on voyages of exploration had themselves tattooed. For many years it was popular only among sailors and soldiers, but many more people now have themselves tattooed. Some people regret it later. It used to be difficult to remove tattoos but lasers are now being used for this.

Legs, arms, body and face are all suitable for tattooing, which is now often done by electric needles. Abstract patterns are very popular, but animals, flowers, plants and even human faces are also reproduced.

In Burma some people cover their entire body with tattoos, and in parts of New Guinea tattooed girls are considered great beauties.

▼ WHY DO HAIRSTYLES VARY SO MUCH?

A person's choice of hairstyle is influenced by fashion and by the type of hair the person has. Types of hair vary. It may be short and crisp, straight and fine, or wavy and curly.

If you want to look fashionable or unusual, the way you style your hair is as important as the way you dress.

Hairstyles also say a lot about your opinions and

customs. In the early 17th century many English men wore their hair down to their shoulders. But the religious Puritans did not approve of this and had their hair cut short. They became known as 'Roundheads'.

Nowadays, hairstyle is mainly a matter of choosing a style you like and that suits your hair. In the 17th and 18th centuries many people wore wigs and hair pieces. These are now used mostly by people with little hair, and by actors who need to change their appearance.

ARTS, SPORT AND ENTERTAINMENT

▲ WHY DO HUMAN BEINGS MAKE MUSIC?

Music is one of the most important ways of giving expression to our thoughts and feelings. Even if no words are sung, it can still be a dramatic way of expressing our emotions. It is quite possible that our prehistoric ancestors made use of music as a way of communicating even before they knew how to speak or write.

Music is a way of blending sounds that are made up of three basic parts – rhythm, harmony and melody. These ingredients may be used on their own, or they may be combined in endless ways. Music may be as simple as a lullaby or as complex as an opera with dozens of singers and instruments all blending together to create sounds of enormous richness and power.

Music can be produced by voices, clapping hands and stamping feet or by musical instruments.

▼ WHY DID HAYDN WRITE THE 'FAREWELL' SYMPHONY?

Franz Josef Haydn (1732–1809) wrote this symphony as a gentle hint to his employer, a member of the noble Hungarian Esterhazy family, that he and the orchestra needed a rest and a vacation. As the music drew to a close, one musician after another set down his instrument and stole away until there were only two violins left.

Haydn is often referred to as the 'father of the symphony'. Although he did not invent this form of music he was the first person to arrange the instruments of an orchestra into four main groups – strings, woodwinds, brass and percussion. He created a big balanced sound in which all instruments had an important part.

Haydn's influence on other musicians of the 18th and 19th centuries was immense. The composer Mozart was a close friend of his and the young Beethoven once studied as his pupil.

Haydn's musical output was enormous. As well as 104 symphonies he also composed innumerable operas, oratorios, masses, concertos, piano sonatas and music for string quartets. One of Haydn's most famous pieces is *The Creation*.

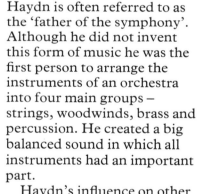

▶ WHY DO BALLERINAS WEAR BLOCKED SHOES?

Blocked shoes, with square solid toes, are used by ballerinas who dance on the tips of their toes, a position known as on *pointe*. The peculiar shape of these shoes makes it easier for them to balance.

It takes considerable skill for a ballerina to do *pointe* work. Girls usually have at least two years of experience before they begin. The first exercises are performed with the help of the barre. Later the dancers train in the centre of the studio floor. Traditionally, only ballerinas dance on *pointe*. Male dancers do not perform this movement.

Originally, all ballet dancers were men. It was not until 1681 that women began to appear regularly on the public stage in Paris. At this time ballet was greatly encouraged by the interest shown by Louis XIV of France. He recognized it as a form of high art and founded the Royal Academy of Dance to train dancers.

▶ WHY DOES AN ORCHESTRA NEED A CONDUCTOR?

Really big orchestras may have a hundred players or more. It takes the control of a skilled conductor to make sure everyone plays in time with each other.

Today, a typically large orchestra has about 90 instruments divided into four sections: the strings, the woodwinds, brass and percussion.

Each instrument usually plays only a part of the entire musical piece, resting from time to time while other instruments take up the music. It is the conductor's task to guide the players in and out of the music at the right moments.

A conductor also rehearses the orchestra before the concert, preparing it to play the music in a way that satisfies him or her. The conductor will teach the orchestra to interpret the music. In the end, it is the conductor who controls the overall sound that the orchestra produces.

▲ WHY DOES AN OBOE PLAY BEFORE THE REST OF AN ORCHESTRA?

The sound of an oboe has a very steady pitch. For this reason, it is the instrument to which all others in an orchestra tune. Before a performance the oboe can be heard playing an A note. It is followed soon after by what sounds like a general groaning and honking noise as the rest of the orchestra gets into tune.

The oboe is a woodwind instrument. Like all woodwinds (except the flute) the sound is produced by blowing through a reed.

Inside the hollow shaft of the oboe a column of air is set vibrating and a sound is created. Holes up and down the various sections of the oboe can be covered with the fingers or with pads. As they are opened and closed, the length of the column of air alters. The longer the column, the deeper the note. The shorter it is, the higher the note that is played.

◀ WHY DO ACTORS WEAR MAKE-UP?

Actors use make-up when they are performing in order to look more 'natural' or else to change their appearance to suit the part they are playing.

Make-up can improve on nature or disguise it entirely. Theatre actors may wear thick make-up to highlight their eyes and lips and features which might otherwise be washed out by the powerful stage lights. For example, a blonde actress can be made to look blonder still if she wears dark make-up on her face. A villain, however, looks more evil if the eyebrows are made heavier and closer.

In films, make-up helps to give the skin a smoother texture and an even tone. This is necessary for close-up shots. Heavier make-up is needed to make actors become wrinkled and grey with age. The most dramatic examples of make-up are found in horror and science-fiction films in which actors portray strange monsters and aliens.

▶ WHY DO PLAYS AND FILMS HAVE DIRECTORS?

The director of a play or film has the same task as the conductor of an orchestra. He or she rehearses the actors and helps them to interpret their parts. Directors must also work from a raw script and assemble from it all the parts that go into making a finished production.

One very important thing is the way a film or play looks to the audience. The director works with designers, lighting and sound crews to make the stage sets look authentic.

A script writer also works closely with the director to change the dialogue or the action if this is needed. The director will also decide which actors are best suited to which parts.

It is important that the director has the production ready on time and on budget. Here he works closely with the producer, who handles the business details and the publicity.

◀ WHY WAS ODYSSEUS TIED TO A MAST?

As he passed their island, Odysseus wished to hear the song of the Sirens. Their music was enchanted and beautiful. It drove sailors to leap from their ships to swim to it, only to be dashed on the rocks and drowned. Odysseus had himself bound to the mast so that he could hear the sound but not fling himself into the sea. The Greek poet Homer wrote this story.

Odysseus' plan was to make his men block their ears with wax so they could hear nothing. He was very curious and wanted to listen. He gave his men firm orders not to untie him.

As the ship sailed near the island, the enchanted song of the Sirens floated across the waves and sent Odysseus into a frenzy. He tried to tear himself away from the mast to hurl himself overboard. His men obeyed his orders and untied his ropes only when the ship had sailed safely away.

▶ WHY DID MICHELANGELO TAKE YEARS TO DO ONE PAINTING?

The scenes that Michelangelo painted in the Sistine Chapel, in Rome, are so huge that it took him four years to complete them.

In 1508 Michelangelo started work on a giant fresco which was to cover the ceiling of the Sistine Chapel. In this type of work, the artist applies paint to wet plaster before it dries.

To reach the high, vaulted ceiling of the chapel, Michelangelo had a tall network of scaffolding built. Every day he would crawl to the top of it, from where he would work lying on his back.

In this uncomfortable position he divided the ceiling into nine panels, each showing a scene from the Old Testament of the Bible, beginning with the Creation.

◀ WHY IS THE MONA LISA SMILING?

The gentle smile of La Gioconda, or Mona Lisa as she is most often known, has made this painting one of Leonardo da Vinci's most famous. While he painted her portrait, Leonardo used musicians, singers and jesters to keep her in a merry mood.

The pleasure reflected in Mona Lisa's face was captured with great skill by Leonardo. The painting seems to be as alive as the beautiful woman who sat for it. Although Leonardo worked on the painting for over four years, he eventually left it unfinished.

Leonardo da Vinci was a man of such extraordinary genius that it is hard to compare him with other men of his time. He seemed to master any subject to which he turned his inquiring mind. His talents are shown in the drawings and paintings he produced, but he also designed cities and drew plans for machines and inventions.

▶ WHY ARE SOME PAINTINGS CALLED ABSTRACT?

Paintings in which the subject is not obvious or in which the artist has made no effort to show real objects are known as abstract paintings. They may contain lines, colours and shapes, but these do not form a recognizable object or person.

Abstract art is one way for a painter to leave aside the appearance of things, and search for the deeper truth that may lie inside them. This might be the difference between showing the beauty of a face and showing that all mankind is mortal.

Abstract art as we know it first appeared before World War I. Since then, it has become very popular. The painting shown here is by the Dutch artist, Piet Mondrian.

Artists often try to express their feelings with abstract art. If a painter tries to show anger with colours and shapes, the end result can be much like a piece of music.

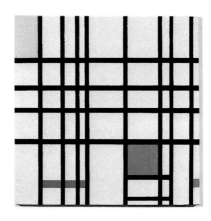

▲ WHY ARE THERE RINGS IN THE CIRCUS?

Circus performers work in a ring because it is easier for them to be watched and for the crowd to see that nothing has been faked.

The standard size of a circus ring is 13 metres in diameter. It has been so ever since the late 18th century when the Franconis, a French circus family, introduced it. But the first person to bring rings into the circus was Philip Astley, who devised the modern circus in 1763. He found that if he galloped in a tight circle while standing on a horse, the centrifugal force helped him to stay upright. The ring that was traced in this way was copied by other trick riders.

The name 'circus' was first used for a spectacular show with riders and acrobats in 1782. The word was borrowed from the ancient Romans who built great oval-shaped stadiums for horse-racing. The biggest, the Circus Maximus in Rome, held 150,000 spectators.

◀ WHY DO CIRCUSES HAVE CLOWNS?

The lure of the circus is its spectacular displays of human skill and daring which amaze and thrill the crowds. When the excitement becomes too great for the audience, it is eased by the clowns. By making the spectators laugh, the clowns help them to relax once more.

Oleg Popov of the Moscow Circus is one of the greatest clowns ever to have worked in the circus, and he is famous in Europe and America as well as the USSR. Unlike many other clowns, who disguise themselves in elaborate floppy costumes and brightly painted faces and wigs, Popov uses much less make-up.

The role he plays is that of the simple bumpkin trying to copy the great riders and acrobats. He impersonates them brilliantly and his efforts to imitate their acts almost seem to succeed. But in the end they are always bungled. In actual fact, Popov is as skilled as some of the stars themselves.

▶ WHY DO TIGHTROPE WALKERS SOMETIMES CARRY LONG POLES?

A long pole helps a tight-rope walker to balance while crossing the rope. It has the same effect as holding out one's arms when walking along a narrow plank or rail.

Not all tightrope artists use poles however. Some of the greatest performed all kinds of amazing feats without ever using them.

Perhaps the best-known was the legendary Blondin (1824–1897), who performed his most daring stunts outdoors. Blondin was the stage name of Jean-Francois Gravelet who astonished the world by crossing Niagara Falls on a tightrope. The rope reached 335 metres from bank to bank at a height of some 49 metres above the water. Blondin first performed this act in 1859 but repeated it several times with variations.

On one occasion he crossed blindfolded, on others he used stilts, pushed a wheelbarrow or carried a man on his back.

▶ WHY DO MAGICIANS USE PROPS?

Stage magicians use props such as tall hats, magic wands, scarves and capes as they practise a kind of magic based on 'sleight of hand' and other conjuring tricks.

Stage magic is only meant for entertainment. If it is well performed it gives the impression that the tricks are truly magical, whereas in fact they are really the clever manipulation of objects.

For instance, a magician may place a bowl of flowers into a box, close the sides carefully and drape a scarf over it. Muttering magic words, he then whisks it away, opens the box and reveals that the flowers have vanished. However, it is not the magic words that make the flowers disappear but rather the box itself. It may have false sides that hide the flowers, or a detachable back or similar device that makes them vanish from sight. The magician uses his props in clever ways to make us believe in his magic.

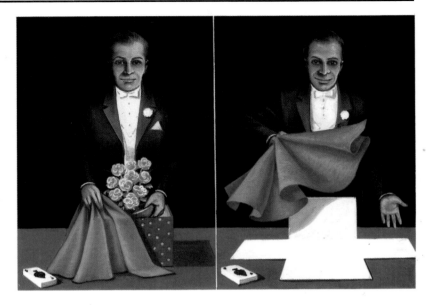

▲ WHY DO PEOPLE BELIEVE IN MAGIC?

In ancient times, tribal witch doctors practised a kind of medicine that mixed together healing herbs, magic charms and strange chants. Often their success was thought to be due as much to their magical powers as to the plants they used. For the people of the time, this was ample proof that magic existed, and that it worked. Many people still believe in magic.

Magic can involve potions, sacred objects and elaborate rituals. Some magic works on the basis of like creating like, so that a rain dance is used to bring rain. Some magic works by contact, so a sacred bone may help to soothe a fever.

There are no physical processes in Nature that could be said to be 'magic'. But the powers claimed by magicians can have such amazing results that we cannot understand them. Possibly magic works because it taps some mysterious sixth sense that we still cannot explain.

◀ WHY ARE THERE KINGS AND QUEENS IN CHESS?

All the pieces in chess take their names from life in the late Middle Ages, when the game first became popular.

Chess is thought to come from India originally, where the pieces used included kings, viziers, chariots, elephants and foot soldiers. It is a game that imitates military life with attacks, retreats, defence and capture. As it spread through Persia, Arabia and finally Europe, the players and rules gradually changed. Today chess pieces include kings, queens, rooks (or castles), knights, bishops and pawns. The two most important and powerful pieces on each side are the king and queen.

The king is the only piece that cannot be captured. It can only be checkmated. This is when the game ends. The term 'checkmate' shows the eastern origins of chess. It comes from the Persian words *shah mat*, which means 'the king is dead'.

► WHY DO SNAKES 'DANCE' FOR SNAKE CHARMERS?

Snakes do not really dance to the music of snake charmers. Their slow, swaying motion is only a way of following the movements of the charmers while they prepare themselves to strike.

Cobras, the snakes most favoured by Indian charmers, do not respond to the music of a pipe. In fact they are deaf to high-pitched sounds. It is the closeness of the charmer and his slow movements that excite a snake and provoke it to rear up into a ready-to-strike position with the hood at its neck flared. Fortunately cobras are rather slow to strike and the charmers know when to stop before the snakes have time to lunge.

Indian cobras are very dangerous however. Their venom is a powerful poison and every year several thousand people die from their bites, usually after the cobras crawl into homes at dusk searching for rats.

◄ WHERE DID PUPPETS COME FROM?

Puppets were known long ago in ancient Greece and Egypt. They were always a very popular form of entertainment and were used to tell legends and folk tales.

In Italy in the 16th century, puppet shows were regular events. The puppeteers travelled far and wide to give performances at fairs and market places. Popular puppet characters soon began to appear. In Italy a hero called Pulcinello was based on a character who was popular in theatres of the time. In England, Pulcinello became known as Punch and by 1800 Punch and Judy shows were found all around the country.

There were also other kinds of puppets. Larger versions that were held on stout rods, and whose heads and arms were moved by smaller rods, were also known. Marionettes consist of an entire human or animal 'body' supported by thin strings from above by puppeteers.

► WHY IS BULLFIGHTING FOUND MAINLY IN SPAIN?

The native wild bulls of Spain were an especially ferocious breed that would fiercely attack humans. Their descendants today supply the bull-rings and make this sport a favourite of Spain and the Spanish-speaking world.

As long ago as Roman times Spanish wild bulls were hunted by men with axes and lances who 'played' with the beasts before killing them.

Modern bullfighting descends from this custom. Today it has more to do with the skill of the matador than with the killing of a bull. The art of a bullfight, or *corrida* as it is known, is the matador's ability to control the bull with his cape. He must be graceful and daring in his movements and be able to thrill the crowd by working as near to the bull's horns as possible.

Even the best matadors are often gored by bulls, and perhaps as many as a third of the greatest have been killed in the ring.

◀ WHY DO PEOPLE READ HOROSCOPES?

The movements of the major heavenly bodies are believed to shape human affairs and to influence human character. Horoscopes help us to understand how these movements affect our everyday lives. For this reason, many people are fascinated by horoscopes and read them every day to find out more about themselves and their future.

A horoscope is the chart used by astrologers. It shows the position of the Sun, Moon and the planets, and also the signs of the Zodiac. A horoscope is divided into sections, or houses, that stand for wealth, happiness, health, friendship, death and so on. As the planets move through the heavens, they travel from one house to the next, influencing each in turn.

Astrologers plot a person's horoscope from the time of birth in order to understand their nature, and to learn about their future.

▶ WHY ARE THERE 52 CARDS IN A PACK?

The number of cards in a modern deck may have links with the lunar year. There are as many cards as there are weeks in the year. The total sum of the cards, counting aces as one, Jacks as 11, Queens as 12 and Kings as 13, is about the same as the number of days in the year.

The exact origin of playing cards is unknown. Most likely the Chinese were the first to use them, since they were the first people to use paper and paper money. The earliest reference to cards comes from the 10th century, during the time of the Liao Dynasty.

Cards appeared in Italy during the 13th century. Early Italian packs had 78 cards. By the 16th century, French packs had become the most widely used in Europe. They were divided into two red suits of hearts and diamonds and two black suits of spades and clubs. Games played with cards include whist, bridge and poker.

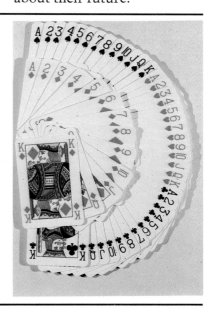

◀ HOW DO VENTRILO-QUISTS 'THROW' THEIR VOICES?

Ventriloquists do not 'throw' their voices; they simply talk without moving their lips. An audience may be fooled into thinking they are not speaking at all and that the voice is coming from somewhere else.

The true secret of a ventriloquist is to fool the audience that not a word is being spoken by him. This is done by forming the words in a normal way but then breathing out very slowly and muffling the sounds by tightening one's throat. The lips must not move at all.

It always helps a ventriloquist if there is another object near him which might just be the source of the voice. A dummy doll, for example, makes a perfect device. Usually the spectators' imaginations will fool them into believing that the dummy is speaking. This is even more likely if its mouth also moves.

▼ WHY ARE THE OLYMPIC GAMES HELD EVERY FOUR YEARS?

When the Olympic Games were revived in 1896, 1500 years after last being held, it was decided to stage them every four years, exactly as had been done in ancient Greek times.

The first recorded Olympic Games were held in 776 BC at Olympia, from where they take their name. They grew out of a religious ceremony held there that involved a sprinting race. According to legend, a Cretan named Herakles proclaimed that the race and ceremony were to be held there every four years.

In time the games became the most magnificent sports festival of the ancient world. Competitors came from far and wide and from many kingdoms and states. It was the custom to arrange a formal truce so that everyone could compete, friend and enemy alike. Today the games serve once again as a kind of truce, as well as bringing nations together.

▲ WHY ARE PROFESSIONAL SPORTSMEN DIFFERENT FROM AMATEURS?

Professionals think of the sports in which they compete as full-time work from which to make a living. Amateurs compete more for love of the sport than for money.

Professional footballers, boxers, tennis players and so on all receive money for their skills. This comes partly as salary and partly from advertising earnings. In the case of amateurs, there are strict rules laid down by athletic associations. These limit what athletes may earn, either from their own performance or by acting as coaches and trainers.

But competition is expensive and amateurs may be helped with training facilities, equipment, food, medical expenses, lodging and transport. They may accept scholarships to colleges which help to support them while they train. Amateurs receive trophies, cups or medals when they win.

▼ WHY ARE VERY LONG RACES KNOWN AS MARATHON EVENTS?

The highlight of long-distance racing in the Olympics is the marathon. It takes its name from a legendary run made by a Greek soldier in 490 BC. He raced from the plain of Marathon to Athens to announce the news of a great victory over an army of invading Persians.

In 1896, at the first modern Olympics in Athens, it was decided to stage a long road-race. Most of the 16 runners in the first marathon were Greek, but they were untrained and inexperienced and most of them soon collapsed with exhaustion. The winner, a Greek called Spiros Louis, was finally joined by members of the Greek royal family who jogged alongside him on the last lap to the finish line. His victory was a sensation.

Since 1908, the marathon has been fixed at 42.2 kilometres, a distance regularly run in about two and a half hours nowadays.

▼ WHY DO RUNNERS USE STARTING BLOCKS?

Starting blocks allow a sprinter to drive forward smoothly and powerfully, and to reach top speed much more quickly than if a standing start was used. The blocks are only needed in sprint races.

Starting blocks were first used in 1927 by the American running coach George Bresnahan. His were made from wood or metal and could be adjusted to support a runner's feet at different angles and distances apart. At Olympia in Greece there are signs of grooves in the track where racers in ancient times might have obtained a good toehold to keep from slipping as the races began.

Until 1884, sprinters started in standing or leaning positions. That year, a Scottish runner began to use a crouched position with one foot in front of the other and both hands touching the ground. From this coiled crouch a runner could spring forward into full racing speed.

▼ WHY DO RUNNERS START FROM DIFFERENT POSITIONS IN SOME TRACK RACES?

When track races are run in lanes, the runners on the outside cover more distance than those on the inside as they come round bends. To ensure that everyone travels the same distance, the runners start at staggered intervals.

Although the 100-metre dash is run in lanes, the race is on the straight and a staggered start is not needed. Longer races of several laps, where the runners have plenty of time to bunch up on the inside lane, also do not need staggered starts.

It is in the 200-metre and 400-metre sprints, in which all the runners travel flat out in different lanes, that staggered starting positions are needed. These races take the runners round several bends. Unless they start well ahead, those on the outside would end up running much farther than runners on the inside.

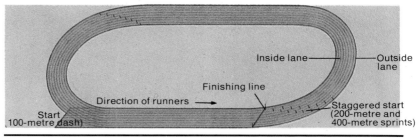

Inside lane — Outside lane

Finishing line

Direction of runners →

Start (100-metre dash)

Staggered start (200-metre and 400-metre sprints)

▲ WHY CAN SOME RACES BE WON BY WALKING?

Walking races are also to be found in the Olympics. The rules are strict, for at no point do they allow the contestants to break into a run to beat their rivals.

Race-walking is also known as heel-and-toe racing, as the racers must never break contact with the ground. One foot must always be in touch. The legs must also be held straight as if walking and not be bent, as when running.

The method for race-walking is nothing like normal walking. The hips are rolled rhythmically, the legs are pulled or jerked up and down very rapidly, and the shoulders and arms are swung in very exaggerated pumping movements. The overall effect is very comical to watch. It is something like a penguin's waddle, yet it allows racers to move along at up to 16 kilometres per hour.

Walking races are very long. In the Olympics the events cover 20-kilometre and 50-kilometre distances.

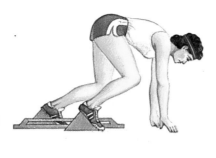

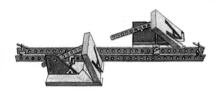

▼ WHY IS DISCUS-THROWING STILL A SPORT?

Throwing the discus was a popular sport in ancient Greece. It was often pictured on vases and praised by poets such as Homer. As a recognition of its past importance, it was revived at the first modern Olympics in 1896.

The discus is shaped like a dinner plate with a thick center that tapers to a thin outer rim. It weighs two kilograms (about 4½ pounds).

The discus is thrown from inside a large boundary circle surrounded by a wire cage on three sides so that a wild shot cannot hit the crowd. The thrower faces backwards as he starts, then whirls rapidly through 1½ turns before releasing the discus. Because of its odd shape, the discus is laid flat against the thrower's palm and held in place by his fingertips. As a result, it is not the accuracy of the throw that counts but the distance that is achieved. Present world records are well over 220 feet.

▼ WHY ARE POLE-VAULTERS NOT HURT WHEN THEY LAND?

A fall from a height of more than 16 feet on to one's back or shoulders could easily break bones. To avoid this, pole-vaulters land in special pits filled with soft rubber pieces or inflated airbags.

Pole-vaulters make a long approach-run to the hurdle to build up speed. As they approach the uprights, they plant one end of the light-

weight fiberglass pole in a sunken take-off box. In a single continuous motion they hurl themselves upwards, swinging feet first towards the bar. As the vaulters continue to pull their feet upward, the right leg crosses over the left leg. The chest is now next to the cross-bar. At the last moment they push the pole away and twist over the cross-bar.

Pole-vaulting arose from the countryside tradition of using poles to jump wide ditches and streams.

▼ WHY ARE THERE BARRIERS IN SOME RACES?

Races in which runners meet barriers across the track are known as hurdles if they are short sprints, and steeplechases if they are long-distance events. The barriers test a runner's ability to clear obstacles without breaking stride or losing speed.

The steeplechase is a 3000-meter event in which each lap of the course has four barriers

cantabrian

and one water jump. The runners either hurdle the barriers or step on to them and leap from the top. The water jump is a barrier followed by a long pool of water. Here, the runners all spring from the top of the barrier to take the next step in the water.

Hurdles races are either 110-meter or 400-meter sprints in which there are ten barriers in all. The obstacles are just over 9 meters apart in the 110-meter race, and 25 meters apart in the 400-meter race.

▼ WHY IS GYMNASTICS AN OLYMPIC SPORT?

Gymnastics was a highly popular activity in ancient Greece. Interest in the sport was revived in the 19th century and it was included in the first modern Olympics in 1896. Women's gymnastics was restricted to only one event until 1952.

Gymnastics consists of a series of various physical exercises that are performed on equipment such as the

horizontal bar, parallel bars, rings, balance beam, pommel horse and vault horse.

In addition there are floor exercises which are performed to music. At the highest levels of competition, these exercises resemble dancers' movements in their graceful style.

For many years the Soviet Union and Eastern European countries have reigned supreme at the Olympics, although since the 1960s the Japanese have produced outstanding performances.

▼ WHY IS REAL TENNIS DIFFERENT FROM TENNIS?

Real tennis, or royal tennis, was first played in the Middle Ages. It is played on an indoor court divided in two by a net. The sloping roof of the court may be used as a surface against which to hit the ball. The game of modern tennis, or lawn tennis, was devised in England in 1873. This game is played on level grass courts or hard-surface courts.

Lawn tennis is highly popular throughout the world. It is a game for two or four players. They use rackets of wood or metal to hit a cloth-covered ball across a long, low net that divides the large open court in half. A singles court for two players is 9 yards wide, whereas a doubles court for four players is $10\frac{1}{2}$ yards wide.

The object of the game is for the players to hit the ball to their opponents so that they cannot return it properly. In this way points are scored.

▼ WHY DO MOST GOLFERS HAVE HANDICAPS?

A handicap is a way of scoring so that a good player can be fairly matched with a weaker player. Each golfer is given minus points, depending on the average number of strokes he or she needs to complete the game. Better golfers take fewer strokes so they have lower handicaps. Weak golfers have higher handicaps because they take more strokes.

The aim of golf is for each player to hit the ball from the starting point, or tee, into a small hole with the least number of strokes. The golfer uses a selection of different clubs. The distance between the tee and the hole can be from 100 to 600 yards. A complete game has 18 holes.

Golf is an ancient Scottish game dating from well before the 15th century. The oldest club still running is the Royal and Ancient Golf Club at St Andrews in Scotland, which was founded in 1754.

▶ WHY IS A YELLOW JERSEY WORN DURING THE TOUR DE FRANCE?

The annual *Tour de France* is the richest and most famous cycle race in the world. It lasts three weeks and is divided into 20 different stages. The winner of each stage wears a much-prized yellow jersey during the next stage of the race.

The *Tour de France* was first held in 1903. It begins each summer in a different French town but always ends in Paris. During the race, the cyclists travel 4,000 kilometres in stages of about 200 kilometres a day. The route varies every year but it takes the race from one side of France to the other. About 12 teams enter the race and each one is heavily sponsored.

▶ WHY DO RACING DRIVERS WEAR FLAME-PROOF CLOTHES?

In a racing accident the greatest danger always is that a fuel tank will rupture and catch fire. Drivers wear flame-proof overalls, boots and gloves that give them about a minute of protection in a burning wreck. They also have a special oxygen supply to their helmets so that they can avoid breathing poisonous fumes from spilled or burning fuel.

The cars driven by Formula One racing drivers are very compactly built. The narrow cockpit is surrounded by fuel tanks. However, the tanks are lined with special self-sealing material that is designed to close any small puncture that may occur in an accident.

The main danger facing all drivers is not that they will be killed in a crash, but that they may be trapped alive in the wreckage while spilled fuel explodes into flames. Their clothes are designed to protect them from the flames while they escape or are rescued.

◀ WHY ARE COLOURED BELTS WORN IN KARATE?

A karate fighter's level of skill is shown by the colour of the belt he or she wears. Masters wear black belts. Students wear brown, blue, green and orange belts, down to white belts for beginners. Students move upwards from one grade to the next by taking formal exams.

Karate, an oriental form of unarmed combat, was first practised in the Ryukyu Islands in the 17th century. Early in the 20th century it spread to Japan and from there throughout the world.

Karate fighters train to focus the entire muscle power of their body into one blow of great force. Hands, fists, elbows and feet are all used to deliver karate blows.

The training for this form of combat is very hard. It involves strengthening the parts of the body that deliver blows as well as practising breathing exercises.

▼ WHY ARE SUMO WRESTLERS GIANTS?

Sumo wrestlers are enormous men. The best fighters stand well over 1.83 metres high and may weigh 130 kilograms or more. The bigger a wrestler is, the greater are his chances of becoming a champion.

Sumo wrestling is enormously popular in Japan. It takes place in a small sand ring surrounded only by a line of small markers. The object of a fight is for one wrestler either to down his opponent by forcing him to the ground or to drive him physically out of the ring.

The wrestlers fight barefoot and almost naked except for a massive belt-like loin-cloth. Their hair is long and tied up in a traditional knot.

A match is usually very short. It begins with a sudden clash as the two giants hurl themselves together. Each wrestler seeks to throw the other off balance and fling him from the ring. Most matches last under a minute.

▼ WHY DO ICE-HOCKEY PLAYERS WEAR MASKS AND PADS?

Goalkeepers are the only players who wear full face masks during ice-hockey games. They give protection from flying pucks and slashing sticks.

The pucks are not only solid rubber but are made rock-hard by being stored in refrigerators before a game. The pucks rocket towards the goal-mouth at speeds up to 160 kilometres per hour.

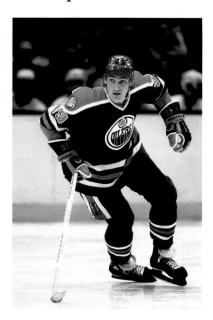

All six players on a hockey team wear pads when they take to the ice to protect their shoulders and knees. This protection is vital because a fall on to the ice at high speed can produce serious injuries. Pucks, sticks and skates can also become lethal instruments in a hard-played match.

The use of protective headgear became common after the death of a professional Canadian player in the 1960s, who hit his head on the ice after a heavy fall.

▼ WHY DO BOXERS WEAR GLOVES?

Padded gloves protect a fighter's hands. They also spare his opponent from the worst cuts and injuries.

In ancient Greek and Roman times, boxers wore weighted leather gloves when they fought. These offered their hands a certain amount of protection but they also made their blows extremely damaging.

The sport had become slightly less savage by the time the first organized bouts were held in Britain in the 18th century. A bare-knuckle style of fighting was used .then. Padded gloves were not commonly used until the late 19th century.

At the 1968 Olympics a new kind of soft leather glove without a surface seam was introduced. Even this small change made a great difference. It reduced serious cuts around the eyes from 46 during the 1964 Games to less than ten in the 1968 Games.

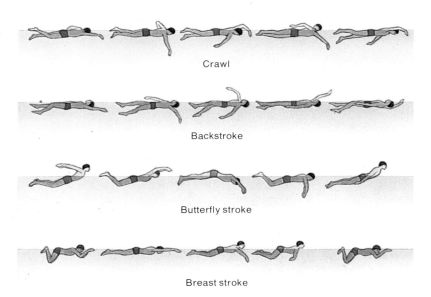

Crawl

Backstroke

Butterfly stroke

Breast stroke

▲ WHY ARE SOME STYLES OF SWIMMING FASTER THAN OTHERS?

Water forms a barrier to the human body as it tries to move through it. Some styles of swimming reduce resistance to the water more than others. This makes them faster.

The most efficient way of travelling through water is a stroke called the crawl. With this stroke, the body cuts through the shallowest amount of water. The crawl is the fastest of all swimming styles and is used by record-setting champions. The crawl was developed in Australia, early in the 20th century, from a stroke used by the people of the South Seas.

Other major styles of swimming include the backstroke, the butterfly stroke and the breast stroke. Breast stroke was used in Europe as long ago as the 16th century. With all of these strokes, the body of a swimmer enters more deeply into the water than with the crawl. Therefore they are slower strokes.

◄ WHY DO SURFBOARDS TRAVEL SO SWIFTLY?

Surfboards skim the surface at speeds much faster than the water may be moving.

Surfing began among the islanders of the South Seas, who used it as a way of skimming over calm lagoons.

They lay on their boards and sculled with their hands or paddled with their feet. More experienced riders began to stand up and balance on waves. They controlled their speed and direction by shifting the weight of their bodies back and forward along the board.

Surfing first became a popular sport in Hawaii, Australia and the west coast of America, where the long Pacific swells made the wave conditions ideal.

► WHY IS SOCCER SO WIDELY PLAYED?

Football is the world's most popular sport, and is played in every continent. The high point of international football is the World Cup, which is held every four years.

The rules of soccer were laid down in Britain in the 1860s when the Football Association was formed to control the sport. It is really a simple game in which two teams of 11 players try to score by kicking a ball into the opposing team's goal-mouth.

Players control the ball with their feet or head, but they are not allowed to touch it with their arms and hands. Only the goalkeeper may handle the ball, and then only inside the penalty area in front of his own goal-mouth. A game is 90 minutes long with a break at half-time.

▶ WHY ARE CAMERAS SO IMPORTANT IN TRACK RACES?

The finish of a race, whether a sprint or a long-distance event, can be such a close-run thing that it is next to impossible for the human eye to tell which runner came first. Fast-action cameras, able to take dozens of pictures a second, can record the instant a runner bursts across the finish line. They make it possible for judges to declare the winner.

Major races must be correctly judged. This is especially true of short events such as 100-metre sprints. Here the difference between the winner and the runner-up may be a matter of only a few fractions of a second. A new world record can be set when only a tenth of a second has been shaved off the existing time.

In order to record such close results, cameras are placed directly along the finish line. They are triggered when the line is crossed, and can detect a difference of centimetres.

▶ WHY ARE SKIS SO LONG?

Long skis are easier to control than short ones. On a steep slope, most of the ski simply slides over the surface. It is the edge of the ski which bites into the snow and allows a skier to change direction and speed. The greater the surface of ski that is in contact with the ground, the easier it is to control.

Skis are made of plastic, metal or, less often nowadays, of wood. They are very flexible so they can ride easily over the bumps and hollows of a hill. A strong set of bindings holds a skier's boot in place and makes it possible to travel at speeds well over 100 kilometres per hour.

Cross-country skis are used for walking rather than sliding. They are very narrow.

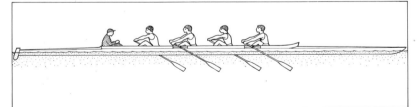

▲ WHY DO ROWING TEAMS OFTEN CARRY PASSENGERS?

Rowing teams of four and eight people are often steered by a coxswain, who at first glance seems to be a passenger. He or she has the very important job of steering the boat and setting the pace of the strokes made by the rowing teams.

Rowing events are usually held in the sheltered waters of rivers and lakes. Racing boats are so light and low-sided that even a small wave could swamp them. They are narrow and very long, and designed to cut quickly through the water.

The rowers sit in single file on sliding seats with their feet braced against stretchers. Each person uses a single oar. These are set in a line on alternate sides of the boat.

INDEX

Page numbers in *italics* refer to pictures

A

Abscess, tooth 40, *40*
Acorn 14, *14*
Actor 106
Adder 31
Aerofoil 54
Aestivation 25
Africa, colonization of 91
Afterburner 55
Air 10, 44, 48
    cooling of 79
    heating of 75, 78
    resistance of 54
Aircraft 54, *54*, 55
Airship 55, *55*
Aix-la-Chapelle 98
Albert, Prince 97
Alcohol 41
Alder 13
Algae 18
Alps 75
Ammonite 72
Andes 75
Animals 18–33
    blood-sucking 20, *20*
    camouflage 32, *32*, 33, *33*
    colouring 32, 33
    eyeless 22
    food 11, *11*
    hibernating 24, *24*
    invisible 32, *32*
    meat-eating 11, *11*
    plant-eating 11, *11*
    as plant food 17
    poisonous 32, *32*
    pollination by 12
    pouched *see* Marsupials
    see also Birds, Fish, Insects
        Mammals, Reptiles
Antarctica 73, *73*
Antelope 30, *30*
Antenna 26, *26*
    false 33
Antibodies 41
Anticyclone 78

Ant 14, 25
    leafcutter 19, *19*
Apollo spacecraft 61
Arabs 93
Ash (tree) 14, *14*
Assisi 98
Asteroids 64, *64*
Astrologer 111, *111*
Astronaut 59, *59*
Athletics 112–14
Atmosphere 63, 66, *66*, 68, 70
    effect on stars 68
    of Venus 63
Axis (Earth's) 71, *71*
Azan 99

B

Babies 39
    baptism of 94, *94*
Bacteria 40, 84
Badger 31
Ball, rubber 44, *44*
Ballast, track, 53, *53*
Ballerina 105, *105*
Ballet 105
Balloon 44, *44*
    hot air 54, *54*
Bandicoot 29
Baptism 94, *94*
Barchan 85
Basalt 84
Bastille 91
Bat 12, 24
    vampire 20, *20*
Beach 77, *77*
Beak 20, *20*
Bean 38
Bear 21, 24
    honey 21
    sun 21, *21*
Beaver 29, *29*
Bee 12, 26
    sting 30, *30*
    swarm 26, *26*
Beech 11
Beetle 22
Belt, karate 116
Beri-beri 38

Berlin Wall 93, *93*
Betelgeuse 67
Bethlehem 97
Bible 95, 96, 97
Bicycle 51, *51*
    see also Cycling
Biplane 54, *54*
Birch 13, 14
Bird-eating spider 30
Birds 11, 12, 14, 24, 25, 28
    courtship of 28
    feeding methods of 20, *20*
    food of 11
    hibernating 24
    mating of 28
    importance in pollination 12
    in seed dispersal 14
    singing of 27, *27*
Blackberry 14, *14*
Black widow spider 27, 30
Bladderwort 17, *17*
Bleeding 36
Blinking 37
Block (railway) 53
Blondin 108
Blood, clotting of 36, *36*
Boats 56
    floating of 46, *46*
    see also Ships
Body
    heat in 48
    make-up of 39, *39*
Boers 91
Bonaparte, Napoleon 91
Bones 38
Boots 103
Boulder clay 83
Bower bird 28, *28*
Boxing 117
Brahma 99
Bread 38
Breathing 39
Bresnahan, George 113
Bridge, railway 52, *52*
Bruise 36, *36*
Bubbles
    air 43, *43*
    soap 43, *43*

Buddhists 99, *99*
Bulb 15, *15*
Bullfighting 110, *110*
Burdock 14, *14*
Burning 48, *48*
Burrow
　shrimp 24
　lungfish 25, *25*
Butterflies 19, 33, *33*
　eyespots of 33, *33*
　proboscis of 19

C
Calcium 38, 39
Calendar, Christian 97
Calories 38
Camera 119, *119*
Camouflage 32, *32*, 33, *33*
　chameleon 33, *33*
　insect 32, *32*
Canals 57, *57*
Cancer, lung 41
Candles 97
Canterbury 98
Carbohydrate 38
Carbon 39
Carbon dioxide 10, 49, 63, 74
Cardboard 43
Cards, playing 111, *111*
Carnivores 11, 24, 31
Carriage, horseless 50
Carrot 38
Cars
　design of 50, *50*
　racing 116, *116*
Castles 89, *89*
Caterpillar 32, *32*, 33
Catfish, electric 20, *20*
Catkin 13
Cave fish 22, *22*
Caves 84, *84*
Cavour, Count 91
Cell 37
　plant 10, 13
Centrifugal force 62
Cereal 13, 38
Ceremonies 94, 95
Ceres 64
Chaffinch 27
Chains, tyres 50
Chalk 84
Chameleon 33, *33*
Charles I, King of Spain 90
Charm, lucky 101
Charon 59

Chartres 98
Cheese 38
Chess 109, *109*
Chinese people 88, 92, 96
Chiron 64
Chlorophyll 10, 11
Christ 95, 96, *96*, 97, *97*
　birth of 97
　crucifixion of 97, *97*
Christian names 94
Christians 89, 95, 96, 97, 98
　persecution of 88, *88*
Christmas 97, *97*
Church, the 89, 94, 95, 96, 97
　medieval 89
Cigarettes 41
Cinnabar moth 32
　caterpillar of 32, *32*
Circus 108
Circus Maximus 108
City-state, Italian 91
Clan 100
Cleaner fish 24, *24*
Cliff 77, *77*
Clothes
　flameproof 116
　need for 35, 48, *48*
　pilgrims' 98
　protective 117
Clouds 78, *78*, 79, *79*
　formation of 79, *79*
Clown 108, *108*
Coal 72, 73, *73*
　in Antarctica 73, *73*
Cobra 31, *31*, 110
　Indian 110
Cocoon 25
Colony 91
　bee 26
　weaver bird 28
Colouring (in animals) 32, 33
Columbus, Christopher 90
Comet 65, *65*
Comet West 65
Communion 95, *95*
Communism 92
Compass 70
Compound, chemical 39
Compressor 58
Computer 45
Conductor
　of heat 48
　lightning 80
　of orchestra 105
Conifer 12
Conjurer *see* Magician

Conquistadores 90
Constantine, Emperor of Rome 88
Constellation 69
Continental drift 73, *73*
Continental plates 73, 74, 75
Corm 15, *15*
Corona 66
Cortes, Hernan 90
Cortex 37, *37*
Cosmetics 101, *101*
　see also Make-up
Couch grass 15
Court, tennis 115
Courtship displays 28, *28*
Coxswain 119
Cnidoblast 18
Crab 23
Crab Nebula 69
Crab spider 32
Cranesbill 12, *12*, 14, *14*
Crater, lunar 61, *61*
Cremation 95
Crevasse 83
Cross-pollination *12*
Crucifixion 97
Crusaders 89, *89*
Crux 69
Crying *see* Tears
Crystal, ice 81, 83
Cuckoo 25, *25*
Curlew 20, *20*
Current
　air 80
　convection 79, *79*
　ocean 76, *76*
　river 82
Cycling 116, *116*
Cyclone 78, 81
Cygnus 69

D
Daisy 14
Dam (beaver's) 29, *29*
Dandelion *12*, 14, *14*
Daphnia 17
Da Vinci, Leonardo 107
Day, cause of 71, *71*
Day of Atonement *see* Yom Kippur
Decay, tooth 40
Decius, Emperor of Rome 88
Decorations, Christmas 97
Deimos 59
Delta 82, *82*

Deposition
    glacial 83, *83*
    river 82, *82*
Desert 85, *85*
Detergent 102
Diet 38, *38*
Dinosaur 72
Diptheria 72
Director
    film 106, *106*
    play 106
Discus 114, *114*
Disease 38, 41
Dormouse 24, *24*
Drainage, natural 82
Dreams 40
Drone 26
Druids 97
Dummy (ventriloquist's) 111

E
Earth (planet) *58*, 59, 70–85
    atmosphere of 68, *68*, 70
    axis of 71, *71*
    crust of 73, 74
    gravity of 42
    heating of 75, 78
    life on 58
    magnetism of 70, *70*
    mantle of 73, 74
    rotation of 71, *71*
Earthquake 74, *74*
'Earthshine' 60
Earthworm 22
Easter 97
Eclipse
    lunar 60, *60*
    solar 61, 66, *66*
Eel, electric 20
Eggs 38
    blow-fly 15
    cuckoo 25
    stickleback 27
    termite 26
Egypt, ancient 101
Electricity 49
Electron 49
Element, chemical 39
Elephant 21, *21*, 87, *87*
Emperor moth 26
Engine
    aircraft 54
    jet 55, *55*
Enzyme 17, 36
Equator 71, 76, 78

Erosion
    glacial 83, *83*
    river 82, *82*
    sea 77, *77*, 84
    soil 85, *85*
Eruption, volcanic 74
Evaporation 47, 76, 82, 84
    from body 35
    of lakes 82
    of sea 76
Excretion 10
Exercise 39
Explorers 90, 91
Eyebright 12, *12*
Eyebrow 37, *37*
Eyelash 37, *37*
Eyelid 37, *37*
Eyes 37, *37*
    colour of 36, *36*
    of fish 22
Eye-spots 33, *33*

F
Factory 90, *90*
Fang, poison 31
'Farewell' Symphony 104
Farming 85, 86
Fat 38, 39
Fern 13, *13*
Ferry 56
Fibrin 36
Fibrinogen 36
Filament (plant) 12, *12*
Films 106
Filter
    photographic 66
    traffic 51
Finch 20, *20*
Fingerprint 34, *34*
Fire, electric 49, *49*
'First-footing' 96
Fish 24, 25, 27, 32
    cave 22, *22*
    cleaner 24, *24*
    electric 20, *20*
    flying 31, *31*
    as food 38
Flames 99
Flatfish 32
Flea 23, *23*
Flowers 12, *12*, 13, *13*
    colours of 12, *12*
    parts of 12, *12*
    stinking 15
Fly 15, 23, *23*
Flying fish 31, *31*

Fog 76, 79, *79*
Follicle, hair 37
Food 10, 11, 38, *38*
    animal 11, *11*
    for hibernation 24
    plant 10
Food chain 11, *11*
Football *see* Soccer
Fossils 72, *72*
    in limestone 84
Fox 11, 31
Franconis, the 108
Freckles 34
'Free fall' 59
Freezing 47, *47*
French Revolution 91, *91*
Frogs
    hibernation of 24
    poison 32
Fruit, fresh 38
Fruits 14
    dispersal of 14, *14*
Funeral 95, *95*

G
Galaxy 69
Ganesh *99*
Garden spider 30
Garibaldi, Giuseppe 91
Gazelle 30
Genes 36
Geotropism 16
Geranium 14
Germicide 102
Germination 16, *16*
Germs 41
Geyser 75, *75*
Giraffe 21, *21*
Glacier 83, *83*
Gland
    lacrimal 37
    sweet 35
Gloves, boxing 117, *117*
Glue 43, *43*
Goby, stargazer 24, *24*
Gods and goddesses 99, *99*
Golf 115, *115*
Gondwanaland 73, *73*
Good Friday 97
Goosegrass 14, *14*
Goose pimples 35, *35*
Grand Canyon 62
Granite 72, *72*
Grasses 13, *13*
Gravity 42, *42*, 64
    effect on plant growth 16

Great Wall of China 88, *88*
Great War *see* World War I
Grebe 28, *28*
Greeks 87, 112
'Greenhouse effect' 63
Growth 10
Gulf Stream 76
Gymnastics 115, *115*

H
Hadj 98
Haemoglobin 38
Hail 80, *80*
Hair 35, *35*
   colour of 37
   grey 37, *37*
   styles 103, *103*
Halley's Comet 65
Halteres 23
Handicap 115
Hannibal 87
Hats 102, *102*
Hawk 11, 20, *20*
Hawthorn 14
Haydn, Franz Joseph 104, *104*
Hazel 13, *13*
Heat 49
   body 35
   conduction of 48, *48*
   effect on flow 46
Hedgehog 24, 31, *31*
Helium 67, 69
Helmet 102, *102*
Henry VIII, King of England 89
Herbivores 11
Heron 20, *20*
Hibernation 24
Himalayas 75
Hindus 98, 99
Hitler, Adolf 93
Hogmanay 96
Homer 114
Homo erectus 86, *86*
Honey 21
Honeybee 30
Hormone, plant 16
Horoscope 111
Horse, Trojan 87, *87*
Housefly 23
House 86
House spider 18
Hoverfly 32, *32*
Humus 84, *84*
Hurdles 114, *114*
Hurricane 78, 81, *81*
Hydrogen 39, 44, 64, 69

Hydrogen sulphide 74
Hypothermia 34

I
Ice 47, 80, 81, 83
   formation of 47, *47*
Iceberg 76
Ice-hockey 117, *117*
Ice-sheet 83
Igneous rock 72
Illness 40
Impala 30
Indians 90
Industrial Revolution 90
Insects 23, 25, 26, 27, 30, 32, 33
   importance to plants 12
   trapped by spiders 18, *18*
   stinging 19, *19*, 30, *30*
   with straws 19, *19*
Instruments, musical 104, 105
Insulators 48
Iron 44, 48, 62
   in food 38, *38*
Iron oxide 44, 62
Islam 98, 99
Israel, founding of 93
Italy, unification of 91

J
Jacobson's organs 23
Jay 27
Jellyfish 18, *18*
Jericho 86
Jersey, yellow (cycling) 116
Jesus *see* Christ
Jewellery 101, *101*
Jews 93, 99
Jupiter 58, *58*, 62, *62*, 64, 68
   satellites of 59

K
Ka'aba 98
Kach 100
Kali 99, *99*
Kangaroo 29, *29*
Kangha 100
Kara 100
Karate 116, *116*
Keel 56
Kes 100
'Killer' bee 30
Kilt 100
Kirpan 100
Knife fish 22
Koala 29
Krait, banded 31

L
Laburnum 14
Lacrimal gland 37
Lakes 82, *82*, 83
   vanishing 82, *82*
Larva
   bee 26
   wasp 19
Last Supper 95
Laurasia *73*
Lava 72, 74, *74*, 84
Leafcutter ant 19, *19*
Leaf insect 32
League of Nations 93
Leather 103
Leaves 11, *11*, 19
Lenin 92
Lent 96
Life
   on Earth 58
   on Mars 62
   on Moon 61
Light 49
   influence on plants 16
   rays 42, 45, 81
   speed of 80
   waves 70
Lightning 80, *80*
Limestone 72, 84, *84*
   composition of 72
Limpet 18, *18*
Lion 32
Liver 38
Lizard 24
Load-line 57, *57*
Lock (canal) 57, *57*
Locomotive, railway 52, *52*, 53, *53*
Lodge (beaver's) 29
'Long March' 92
Longship 88, *88*
Lords and ladies (plant) 15, *15*
Louis XIV, King of France 105
Luminosity of stars 68
Lungfish 25, *25*
Luther, Martin 89, *89*, 97

M
Magic 109
Magician 109, *109*
Magma 74, *74*
Magnetic field (Earth's) 69
Magnitude (of stars) 68
Maize 13
Make-up 106, *106*
   see also Cosmetics

Mamba 31
Mammals 11, 20, 24
    see also Animals
Man, early 86
Mantid 27
Mantle (Earth's) 73, 74
Mao Tse-tung 92
Maple 14, *14*
Marble 72
Marathon 112
Marionette 110, *110*
Marriage 94, *94*
Mars *58*, *62*, 64, 65, 68
    colour of 62
    life on 62
    satellites of 69
Marsupials 12, 29, *29*
Matador 110
Matches 49, *49*
Meander 82, *82*
Measles 41
Meat 38
Mecca 98, 99
Meditation 99
Mediterranean Sea 73, 76
Melanin 34
Melanocytes 34
Mercury (metal) 40
Mercury (planet) *58*, 59, 63, *63*
Metal 48, *48*
    precious 101
Metamorphic rock 72
Meteor 65, *65*
Meteoroid 65, *65*
Michelangelo 107, *107*
Middle Ages 89
Mihrab 98
Milk 38
Milky Way 69, *69*
Mimosa 16, *16*
Minaret 99, *99*
Minerals
    in food 38
    in soil 84
Mirror 42, *42*
Mohammed 98, 99
Molecules 44, 46, 47, 48, 49, 80
    metal 49
    oxygen 48, 49
    rubber 44
    water 43, 46, 47
    wood 48, 49
Mona Lisa 107, *107*
Mondrian, Piet 107
Monoplane 54
Moon 59, 60, *60*, 61, *61*

axis of 60, *60*
craters on 61, *61*
eclipse of 60, *60*
gravity of 42
life on 61
New 61, *61*
phases of 60, *60*
Moons 59, *59*
Moraine 83
Mosque 98, 99, *99*
Moss 13, *13*
Moths 32, *32*, 33
    antennae of 26, *26*
    proboscis of 19
    silk 26
    smell of 26
Mountains 75, *75*
Muezzin 99
Muscles 35, 39
Music 104, *104*, 105
Muslims 89, 98, 99, 100

N
Napoleon Bonaparte 91
Nebula 69
Nectar 12, 19
Needle
    compass 70
    pine 11, *11*
Nematocyst 18, *18*
Neptune *58*, 64
Nero, Emperor of Rome 88
Nests
    cuckoo 25, *25*
    leafcutter ant 19
    termite 26
    stickleback 27, *27*
    weaver bird 28, *28*
New Moon 61, *61*
New World, the 90
New Year 96
    Chinese 96, *96*
    Jewish 96
Newspapers 43
Nicholas II, Czar of Russia 92
Nicotine 41
Night, cause of 71, *71*
Nile fish 22, *22*
Nitrogen 39
North Pole 71
North Sea 73
Numbers 45
Numerals, Roman 45
Nut 14

O
Oboe 105
Ocean *see* Sea
Odysseus 106, *106*
Oil, fragrant 102
Old man's beard 14, *14*
Olm 22
Olympia 112, 113
Olympic Games 112, *112*, 113,
    114, 115, 117
Orbit *59*
    of comet 65, *65*
    of meteoroid 65, *65*
    of Moon 60
Orchestra 105, *105*
    instruments of 104
Orion 67
Ovary 12, *12*
Ovule 12, *12*
Owl 11
Oxygen 39, 44, 48, 49

P
Painting, abstract 107, *107*
Palestine 93
Pangaea 73, *73*
Paper 43, *43*
Parasite 23, 24, 31
Parvati 99
Passover 98, *99*
Pea 12, 14, *14*
Penny farthing 51
Penumbra 60, 67
Perfume 102, *102*
Perseids 65
Petal 12, *12*
Pharaoh 87
Phases, lunar 60, *60*
Phobos 59
Phosphorus 39
Photosynthesis 10, 16
Pigment 34
    in hair 37
    in plants 10
Pilgrimage 98
Pine 11
Pit viper 23, 31
Pizarro, Francisco 90
Placenta 29
Plaid 100
Plane *see* Aircraft
Planet 58, *58*, 59, *59*, 62, *62*, *63*,
    64, *64*
Plantain 13

Plants
  carnivorous 17, *17*
  colour of 10
  flowers of 12, *12*, 13, *13*
  food of 10, 17
  fruits of 14, *14*
  growth of 16, *16*
  insect-eating 17, *17*
  leaves of 11
  pollination of 12, 13
  roots of 16, *16*
  stems of 16, *16*
  stinking 15
  water in 10
Plaque 40
Plates, continental 73, 74, 75
Plays 106
Plimsoll, Samuel 57
Plimsoll Line 57, *57*
Pluto *58*, 59, 65
Poison, snake 31
Poison frog *32*
Poles (Earth's) 70, 71
Pole-vaulting 114, *114*
Police 34, 101
Polio 41
Pollen 12, 13, *13*, 15
Pollination 12, 13
Poorwill 24
Popov, Oleg 108
Potato 15
Pouch (animal) *see* Marsupials
Predator 11
Pressure, air 78, *78*
Primrose 12
Proboscis 19, *19*
Propeller, aircraft 54, *54*
Protea 12, *12*
Protein 38
Puck, hockey 117
Puddle, drying of 47, *47*
Puff adder 31
Punch and Judy 110
Puppet 110
Pyramid 87, *87*
Pyrenees 75

Q
Queen bee 26
Queen termite 26, *26*

R
Rabbit 11
Racing driver 116
Racket, tennis 115

Radula 18
Rafflesia 15, *15*
Rails 53, *53*
  welded 52
Railways
  electrified 53, *53*
  maintenance of 52
  underground 52, *52*
Rain 79, *79*
  effect on soil 85
Rainbow 81, *81*
Rattlesnake 23, 31
Ray, electric 20
Real tennis 115
Receptacle 12, *12*
Red giant 67, 69
Red Sea 73
Reformation 89
Refraction 81
Religion 88, 89, 94–100
Reptiles 23, 72
Resilin 23
Resolutions, New Year 96
Respiration 10
Rest 39, *39*
Revolution
  French 91, *91*
  Industrial 90
  Russian 92, *92*
Rhizome 15, *15*
Ribwort *13*
Rice 13
Richard the Lion Heart, King of
  England 89
Ripples, sand 77, *77*
Rivers 82, *82*
Rock
  igneous 72
  metamorphic 72
  sedimentary 72, 84
Rockies 75
Rodent 12, 14
Romans 85, 88, 93, 97, 108
  circuses 108
  war against Hannibal 87
Rome 98
Roots 16
Rowing 119, *119*
Royal Academy of Dance 105
Royal tennis *see* Real tennis
Rubber 44
Running 113
Rushes 13
Russian Revolution 92, *92*
Rust 44, *44*, 62
Rye grass *13*

S
Sabbath 95
Safety matches 49
Sails 56
Salamander 24
Salt 76
Sand dune 85, *85*
Sandstone 72, *72*
Sand wasp 19, *19*
Santiago 98
Saracens 89
Satellite 59
Saturn *58*, 62
  rings of 64, *64*
  satellites of 59
Saxons 97
Scots 100
Scurvy 38
Sea 70
  currents in 76, *76*
  disappearing 73
  erosion by 77, *77*
  salt in 76
Sea anemone 18
Seasons 71, *71*
Seat belt 51, *51*
Seaweed 18
Sedge 13
  common *13*
Sedimentary rock 72, 84
Seeds 14
  germination of 16, *16*
  production of *12*
Self-pollination 12
Sepal 12, *12*
Setae 22
Sheetweb spider 18
Shellfish 18
Ships
  cargo 56
  container 56
  loading of 57
  merchant 56, *56*
Shiva 99, *99*
Shivering 35
Shoes 103, *103*
  blocked 105
Shrimp, *Alpheus* 24, *24*
Shrine 98
Signals
  railway 53, *53*
  traffic 51, *51*
Sikhs 100, *100*
Silk moth 26
Sirens, the 106
Sistine Chapel 107, *107*

Skater, ice 47
Skin
  bruising of 36, *36*
  colouring of 34, *34*
Skis 119, *119*
Sleep *39*
  importance of 39
  types of 40, *40*
Sleeper, railway 53, *53*
Snail 11
  hibernation of 24
Snake charmer 110, *110*
Snakes 24, 110
  dangerous 31, *31*
  tongues of 23, *23*
Snow 81, 83
Snowdrop 14
Snowflake 81, *81*
Soap 102, *102*
  bubble 43, *43*
Soccer 118, *118*
Sodium chloride *see* Salt
Soil 84, *84*, 85, *85*
Solar System 58, *58*, 59, 64
Sound, speed of 80
South Pole 71
Squirrel 14
Space 58–69
Spacecraft 59
Spain 90, 110
Spiders 18, *18*, 23, 27, *27*
  poisonous 30, *30*
Spines, hedgehog 31
Spinning (top) 45
Spokes 51
Spores 13, *13*
Sportsmen 112
Springbok 30, *30*
Sprinting 113, 119
Squid 33, *33*
Squirting cucumber 14, *14*
Stalactite 84, *84*
Stalagmite 84, *84*
Stapeliad 15
Starch 38
Starlight 59
Stars 58, 67, 68, *68*, 69, *69*
  brightness of 68, *68*
  in daytime 68, *68*
  death of 69, *69*
  formation of 69, *69*
  twinkling of 68, *68*
Starting block 113, *113*
Steel 44
Steeplechase 114
Stems 16

Stick insect 32
Stickleback 27, *27*
Stigma 12, *12*
Sting
  bee 30, *30*
  wasp 19
Stomata 10
Storm 80, *80*, 81
Strokes (swimming) 118, *118*
Style (flower) 12, *12*
Submarine 57, *57*
Submersible 57
Sugar 46, *46*
  in food 38
Sulphur 39
Sulphur dioxide 74
Sumo wrestler 117, *117*
Sun 58, 59, 60, 66, *66*, 67, *67*, 68
  eclipse of 66, *66*
  setting of 66, *66*
  size of 67, *67*
  surface of 66
  effect on water of 47, *47*
Sundew 17, *17*
Sunlight 34, 81
Sunspot 66, 67, *67*
Suntan 34, *34*
Supernova 69
Surface, wet 45, *45*
Surface tension 43
Surfing 118, *118*
Swallow-hole 84, *84*
Sweat 35
  gland 35
  pore 34
Sweet violet 14
Swimming 118, *118*
Symbiosis 24
Syrup, flow of 46, *46*

T
Tanker, oil 56, *56*
Tarantula 30
Tartan 100, *100*
Tattoo 103, *103*
Tawaf 98
Tears 37, *37*
Tee, golf 115
Teeth 38, 40, *40*
  care of 40
Temperature 47, 49
  body 40, *40*
Tennis 115, *115*
Termite 26, *26*
Tetanus 41
Tethys 73, *73*

Thermometer 40, *40*
Thistle 14
'Three Estates' 91
Thrombin 36
Thunder 80, *80*
Tide 76, *76*, 81
Tightrope walker 108, *108*
Timothy grass *13*
Titan *59*
Tobacco 41
Tongue, snake's 23, *23*
Toothache 40
Top, spinning 45, *45*
Touch-me-not 14, *14*
Tour de France 116, *116*
Towns 86, *86*
Toxoid 41
Track
  race 113, *113*
  railways 52; *see also* Rails
Track races 119, *119*
  see also Athletics, Hurdles,
  Running, Starting block,
  Steeplechase
Tractor 50, *50*
Train
  electric 53, *53*
  underground 52, *52*
Transit (planet) 63
Transpiration 10
Transport 50–7
Trees 11, 13
  Christmas 97
Trench warfare 92, *92*
Trojans 87
Tropics 71
Trunk (elephant's) 21
Tuber 15
Tunnel, railway 52
Turban 100, *100*
Turbo-jet *see* Engine, Jet
Turbo-prop 54
Typhoon 81
Tyre 50, *50*

U
Umbra 67
Uniform 101, *101*
United Nations 93, *93*
Upthrust 46, *46*
Uranus 58, *58*
Urban II, Pope 89

V
Vaccination 41, *41*
Vampire bat 20, *20*

Vapour, water 47, *47*, 74, 70
Vegetables 38
Veil 100
Ventriloquist 111, *111*
Venus *58*, 59, 63, *63*
　　atmosphere of 63
　　rotation of 58
　　temperature of 63
Venus fly trap 17, *17*
Victor Emmanuel, King of Italy
　　91
Victoria, Queen of England 97
Vikings 88, *88*
Viking lander 62, *62*
Viking spacecraft 62
Villages 86
Viper 23, 31
Vishnu 99
Vitamin 38, *38*
Voice, throwing of 111
Volcano 74, *74*
Vole 11

W
Walking 86, *86*, 113, *113*
Waning (Moon) 60, *60*
Wasp 19, *19*, 32
　　sand 19, *19*
Water 39, 44
　　flow of 46, *46*
　　freezing of 47, *47*
　　in plants 10
　　in puddles 47
　　upthrust of 46, *46*
Waves 77, *77*
Waxing (Moon) 60, *60*
Weather 78–81
Weaver bird 28, *28*
Web, spider's, 18, *18*
Wedding *see* Marriage
Wegener, Alfred 73
Weight (in space) 59, *59*
Weightlessness 59
Wheat 13
Wheel
　　bicycle 51, *51*
　　car 50, *50*
　　tractor 50, *50*
　　train 53, *53*
Whistling thorn 25, *25*
Whooping cough 41
Widgeon 20, *20*
Wigs 103
Willow 13, 14
Willow herb 14
Willow warbler 20, *20*

Wind 81, 85
　　cause of 78, *78*
　　pattern of 78, *78*
　　for sailing ships 56
　　effect on sea 76
　　in seed dispersal 14
Witch doctor 109
Wood, burning of 48, 49
World War I 92, *92*
World War II 93
Worm *see* Earthworm
Worship 95, 98, 99
Wrestling 117, *117*

Y
Yacht 56, *56*
Yellowhammer 27
Yoga 99
Yom Kippur 96

Z
Zodiac, signs of 111

PHOTOGRAPHIC ACKNOWLEDGEMENTS

Pages: 24 bottom Seaphot/Richard Chesher, 38 All-sport/Steve Powell, 39 Zefa/Corneel Voigt, 41 top Zefa/D. Davies, 41 bottom J. Allan Cash, 43 top Paul Brierley, 43 bottom Gene Cox, 44 J. Allan Cash, 51 J. Allan Cash, 53 top British Rail, 59, 62 & 63 Nasa, 67 centre Lick Observatory, 69 centre California Institute of Technology, 69 bottom Observatorium, Lund, Sweden, 70 Nasa, 72 Pat Morris, 73 British Antarctic Survey, 78 Nasa, 81 centre J. Allan Cash, 81 top Zefa/Sauer, 82 Nasa, 85 J. Allan Cash, 92 Novosti Press Agency, 93 top United Nations, 93 bottom left Popperfoto, 93 bottom right J. Allan Cash, 98 MEPhA/A. Duncan, 103 Picturepoint, 105 Zefa/Starfoto, 107 centre Giraudon, 107 bottom Tate Gallery, 122 Colorsport, 114 left Colorsport, 114 centre and right All-Sport/Tony Duffy, 115 All-Sport/Steve Powell, 116 top All-Sport, 166 centre Colorsport, 116 bottom Zefa/H. Helmlinger, 117 All-Sport, 118 bottom Colorsport, 118 right All-Sport, 119 Colorsport.

Picture Research: Penny Warn and Jackie Cookson.